HEARTBEAT ART

Leonardo

Seán Cubitt, Editor-in-Chief

Social Media Archeology and Poetics, edited by Judy Malloy, 2016
Practicable: From Participation to Interaction in Contemporary Art, edited by Samuel Bianchini and Erik Verhagen, 2016
Machine Art in the Twentieth Century, Andreas Broeckmann, 2016
Here/There: Telepresence, Touch, and Art at the Interface, Kris Paulsen, 2017
Voicetracks: Attuning to Voice in Media and the Arts, Norie Neumark, 2017
Ecstatic Worlds: Media, Utopias, Ecologies, Janine Marchessault, 2017
Interface as Utopia: The Media Art and Activism of Fred Forest, Michael F. Leruth, 2017
Making Sense: Cognition, Computing, Art, and Embodiment, Simon Penny, 2017
Weather as Medium: Toward a Meteorological Art, Janine Randerson, 2018
Laboratory Lifestyles: The Construction of Scientific Fictions, edited by Sandra Kaji-O'Grady, Chris L. Smith, and Russell Hughes, 2018
Invisible Colors: The Arts of the Atomic Age, Gabrielle Decamous, 2018
Virtual Menageries: Animals as Mediators in Network Cultures, Jody Berland, 2019
From Fingers to Digits: An Artificial Aesthetic, Ernest Edmonds and Margaret A. Boden, 2019
MATERIAL WITNESS: Media, Forensics, Evidence, Susan Schuppli, 2020
Tactics of Interfacing: Encoding Affect in Art and Technology, Ksenia Fedorova, 2020
Giving Bodies Back to Data: Image-Makers, Bricolage, and Reinvention in Magnetic Resonance Technology, Silvia Casini, 2021
A Biography of the Pixel, Alvy Ray Smith, 2021
Living Books: Experiments in the Posthumanities, Janneke Adema, 2021
Art in the Age of Learning Machines, Sofian Audry, 2021
Design in Motion: Film Experiments at the Bauhaus, Laura Frahm, 2022
Northern Sparks: Innovation, Technology and the Arts in Canada from Expo '67 to the Internet Age, Michael Century, 2022
The Artwork as a Living System: Christa Sommerer and Laurent Mignonneau 1991–2022, edited by Karin Ohlenschläger, Peter Weibel, and Alfred Weidinger, 2023
Computational Formalism: Art History and Machine Learning, Amanda Wasielewski, 2023

Picture Research: The Work of Intermediation from Pre-Photography to Post-Digitization, Nina Lager Vestberg, 2023
Art + DIY Electronics, Garnet Hertz, 2023
Voidopolis, Kat Mustatea, 2023
After Eating: Metabolizing the Arts, Lindsay Kelly, 2023
Tactical Publishing: Using Senses, Software, and Archives in the 21st Century, Alessandro Ludovico, 2023
The Future Is Present: Art, Technology, and the Work of Mobile Image, Philip Glahn and Cary Levine, 2024
Living Surfaces: Images, Plants, and Environments of Media, Jussi Parikka and Abelardo Gil-Fournier, 2024
Heartbeat Art, Claudia Arozqueta, 2025

See http://mitpress.mit.edu for a complete list of titles in this series.

HEARTBEAT ART

CLAUDIA AROZQUETA

THE MIT PRESS
CAMBRIDGE, MASSACHUSETTS
LONDON, ENGLAND

The MIT Press
Massachusetts Institute of Technology
77 Massachusetts Avenue, Cambridge, MA 02139
mitpress.mit.edu

The MIT Press would like to thank the anonymous peer reviewers who provided comments on drafts of this book. The generous work of academic experts is essential for establishing the authority and quality of our publications. We acknowledge with gratitude the contributions of these otherwise uncredited readers.

This book was set in Arnhem Pro and Frank New by Westchester Publishing Services. Printed and bound in the United States of America.

Library of Congress Cataloging-in-Publication Data

Names: Arozqueta, Claudia, author.
Title: Heartbeat art / Claudia Arozqueta.
Other titles: Heartbeats in the arts
Description: Cambridge, Massachusetts : The MIT Press, [2025] | Series: Leonardo | Revision of the author's thesis (doctoral)—University of New South Wales, 2021, under the title: Heartbeats in the arts : the pulse of history. | Includes bibliographical references and index.
Identifiers: LCCN 2024023057 (print) | LCCN 2024023058 (ebook) | ISBN 9780262551885 (paperback) | ISBN 9780262382601 (pdf) | ISBN 9780262382618 (epub)
Subjects: LCSH: Heart beat in art. | Art, Modern—20th century—Themes, motives. | Art, Modern—21st century—Themes, motives.
Classification: LCC N8217.H53 A76 2025 (print) | LCC N8217.H53 (ebook) | DDC 701/.04561--dc23/eng/20241230
LC record available at https://lccn.loc.gov/2024023057
LC ebook record available at https://lccn.loc.gov/2024023058

10 9 8 7 6 5 4 3 2 1

EU product safety and compliance information contact is:
mitp-eu-gpsr@mit.edu

CONTENTS

SERIES FOREWORD (ix)
ACKNOWLEDGMENTS (xi)

INTRODUCTION (1)

1 MEASURING (13)

Feeling Pulses: Indian, Chinese, Egyptian, Greek Medicine
Listening Heartbeats: René Laennec's Stethoscope
Graphing Heartbeats: Étienne-Jules Marey's Machines
Electrical Heartbeats: Willem Einthoven's Electrocardiograph
Mechanical Heartbeats: Futurism, Weimar Art, Surrealism

2 SPACE (69)

Spatial Data: Laika, Moon Landing
Golden Record: Ann Druyan, Sci-Fi Films, Pulsing Stars
Beat Music: Sun Ra, Milford Graves, Pink Floyd, Kraftwerk
Moon Bounces: Pauline Oliveros, Alvin Lucier
Universe's Vibrating Symphony: Éliane Radigue

3 ELECTRICAL (99)

Vibrating Lines: Heinz Mack
The Visual Energy of Heartbeats: Jean Dupuy
Lighting Bodily Energies: Mark Boyle, Joan Hills
Electro-Trances: Diana Domingues

4 **TIME (137)**

Timing Pulses: Allan Kaprow
Heartbeat Time: Julieta Aranda
The Vitality of Spaces: Sasaki

5 **TRANSPLANTS (159)**

Self-Transplants: Brian O'Doherty
Co-mingle: Catherine Richards

6 **SURVEILLANCE (181)**

Body Information: Teresa Burga
Intrusions: Mona Hatoum
Biometrics: Rafael Lozano-Hemmer

7 **REGENERATIONS (217)**

Getting Closer: Yoko Ono
Heart Endurance: Linda Montano
(Im)Permanence: Christian Boltanski

CONCLUSION (249)

NOTES (253)
INDEX (281)

SERIES FOREWORD

Leonardo/The International Society for the Arts, Sciences and Technology fosters transformation at the nexus of art, science, and technology because complex problems require creative solutions. The Leonardo Book Series shares these aims of artistic and scientific experimentation, and publishes books to define problems and discover solutions, to critique old knowledge and create the new.

In the early twentieth century, the arts and sciences seemed to interact instinctively. Modern art and modern poetry were automatically associated with relativity and quantum physics, as if the two were expressions of a single Zeitgeist. At the end of the Second World War, once again it seemed perfectly clear that avant-garde artists, architects, and social planners would join cyberneticists and information theorists to address the problems of the new world order and to create new ways of depicting and understanding its complexity through shared experiences of elegance and experiment. Throughout the twentieth century, the modern constantly mixed art and science.

In the twenty-first century, though, we are no longer modern but contemporary, and now the wedge between art and science that C. P. Snow saw emerging in the 1950s has turned into a culture war. Governments prefer science to arts education, yet stand accused of ignoring or manipulating science. The arts struggle to justify themselves in terms of economic or communicative efficiency that devalues their highest aspirations. And yet never before have artists, scientists, and technologists worked together so closely to create individual and collective works of cultural power and intellectual grace. Leonardo looks beyond predicting dangers and challenges, beyond even planning for the unpredictable. The series publishes books that are both timely and of enduring value—books that address the perils of our time, while also exploring new forms of beauty and understanding.

Seán Cubitt
Editor-in-Chief, Leonardo Book Series

Roger F. Malina
Executive Editor, Leonardo

ACKNOWLEDGMENTS

Writing a book is a colossal challenge that involves a lot of time and the support of many people and institutions. *Heartbeat Art* has been in the writing for over a decade. It crystallized in its current form from my PhD dissertation. I convey my gratitude to the University of New South Wales for funding my doctoral studies through the University International Postgraduate Award. I am also grateful for other grants awarded to this research, including a Medibank International Fieldwork Grant, a Ross Steele Residency Scholarship for International Exchange, an Arc Student Grant, and a HDR Scholarship of the Australian Government. I would also like to express my gratitude to Douglas Kahn for his invaluable insights and support. Thanks also to Lindsay Kelley for her contributions to this project.

I am grateful to all the artists studied in this book, many of whom I had the opportunity to interview or exchange emails with and who kindly provided images for this publication. My gratitude also to the persons, archives, and institutions who kindly contributed information and images, all of which are credited in the image captions throughout the book.

My interest in heartbeats, art, and science started during my MA in media art histories in Austria in 2010. Since then, I have written some papers and a thesis on the topic. Some parts of the material in this book appear in different forms in the following publications: Arozqueta, C., "Heartbeats in the Arts," PhD Thesis, University of New South Wales, 2021; Arozqueta, C., "Heartbeats and Digital Arts: A Historical Connection," *Leonardo* 51, no. 1 (2018): 33–39; Arozqueta, C., "Connecting Realities: Rafael Lozano-Hemmer's Pulse-Based Works," *Journal of Science and Technology of the Arts* 6, no. 2 (2014): 21–28; and Arozqueta, C., "Seeing the Unseen: Rafael Lozano-Hemmer's Pulse-Based Works," MA thesis, Donau-Universität Krems, 2013. Some papers were also presented at the following conferences: Arozqueta,

C., "Illuminating Bodily Energies," at the Annual Meeting of the Society for Literature, Science, and the Arts, Ann Arbor, Michigan, 2021; Arozqueta, C., "Son et Lumiere: The Body as a Performance Instrument," at the Media Art Histories Conference, Aalborg, Denmark, 2019; Arozqueta, C., "Cardio-portrayals: The Ultimate Abstraction," HDR Conference, University of New South Wales, Sydney, 2017.

I am grateful to Leonardo Book Series editor Seán Cubitt, acquisitions editor Gabriela Bueno Gibbs at the MIT Press, the anonymous reviewers for their invaluable insights, and the MIT editorial team for their care and professionalism. I want to thank Peter Blamey for his proofreading and indexing of this book.

My warmest gratitude to my friends, Mary Barr and Jim Barr, for their friendship, support, and editing of some draft versions of these texts.

I would also like to thank my family, especially my mother, Lupita Arozqueta, for her steadfast love and support.

Finally, my heartfelt gratitude to Rodrigo Azaola, my husband, who listened to me talk about this book for many years and whose love, support, and encouragement were a pulse of motivation throughout various stages of my long research journey.

The majority of this manuscript was written in Sydney/Warang on the lands of the Gadigal people. However, some sections were written in Mexico on the lands of the Mayas, and during research residencies in Paris, at the Cité Internationale des Arts with the support of the Ross Steele Residency Scholarship, and in Montreal at the Conseil des arts et des lettres du Québec Residency, with the support of the Mexican Ministry of Culture.

INTRODUCTION

Knowledge is like the heart, hidden and beating, glowing imperceptibly, regulating the rhythmic canals that flow back and forth and flow into other canals, torrents and unexpected currents managed by radical complexity of a powerful central valve.
—Emilio Carballido, "Yo también hablo de la rosa," 1965

The heartbeat is more than a purely physical phenomenon. Generally defined as the movement or sound of the heart as it sends blood around the body and as a vital or animating unifying force, the heartbeat is also a vibration and bioelectricity. The heart's pumping action is produced by electrical signals that start at its sinoatrial node and later travel through the heart, triggering the contraction of the atria, followed then by the ventricles. The record produced by an electrocardiograph (EKG), called an electrocardiogram, is the trace of that energy translated into a graph.

As a signal of the heart, an organ that has been historically and ubiquitously conceived as the house of emotions, heartbeats are also related to sentiments. In the ancient Sanskrit texts of Ayurveda, the heart is described as the seat of consciousness and the center of the body system that puts in motion prana, the energy that pulses around the body.[1] The Egyptians considered the heart the source of wisdom, emotions, memory, and the key to eternity.[2] In ancient Chinese medicine, the heart was conceived as the emperor of the body that controlled other parts, including the mind.[3] In the Americas, Nahuas and Mayas, through the practice of sacrifice, knew well that the heart's pulsations were the support of life, so they envisioned the heart (*yóllotl* and *puczikal*, respectively) as a vital force that beats, sustains life, and marks an existence.[4] Both Nahuas and Mayas believed that within the precious organ lay an animate entity or energy, the sensitivity, perception, knowledge, and personality of a person.[5] In Europe, in the fourth century BCE, Aristotle thought that the heart was the core of vitality and considered its central location in the body to be intimately related to its key role in its functionality, believing that as a heat organ, it affected emotions; this was in opposition to the brain, which was by its nature immune to feelings.[6]

The heartbeat has been historically linked to emotions and has served as analogy, symbol, and synonym of many feelings, states, and stages of life. Just think about how your own heartbeat was the first sign of your own existence when you were a baby in your mother's womb and how it is a sign that moves every day in tune with your feelings. Your heartbeat flutters when you are afraid or angry. It goes into a big crescendo when you are close to someone you are physically attracted to. The pulse (the rhythmic beat felt in the artery that is an indicator of the blood-pumping action of the heart) decreases when one feels depressed, and it vanishes in the absence of life. Although the brain is known to physiologically suppress the perception of your own heartbeat in order to avoid interfering with the brain's perception of the external world,[7] the heartbeat speaks for itself throughout the body, not only in the circulatory system but also in the circle of emotions. Thus, in the heartbeat, the mind and body converge because the beating of the heart is the origin of life but also of activities, emotions, movements, or ideas. That is the source of the expression "the heartbeat of something," such as "the heartbeat of philosophical discussions" or "the heartbeat of a community," which underscores an important feature that is responsible for making it what it is.

Heartbeats produce a "lub-dub" sound due to turbulence caused by the closure of the heart valves. The "lub" noise marks the start of the systole (the phase of the heartbeat when the heart muscle contracts and pumps blood from the chambers into the arteries), heard when the atrioventricular valves close, and the "dub" marks the end of the systole and is produced by the closure of the semilunar valves. Within these sounds, we can hear murmurs, variations of sounds in the heart cycle, which can express the inner state of the body. Heart sounds are the language of the heart and are also a metric, and, as such, artists have been using them in the arts since early times. For instance, in the sixth century, the music of the spheres was connected to the body in Boethius's second classification of *musica humana*. This type of music was determined by the many identifiable patterns of breath and heartbeats.[8]

Heartbeats are energy, music, graphs, and emotions, and through current biometric and artificial intelligence (AI) technologies, heartbeats are also an identification system, a password contained within a graph, which makes them a signature and profile of a person that can be used for classification, detection, and commercialization purposes. Furthermore, due to advancements in biomedicine, the heart and its precious heartbeats are transplantable and, thus, subject to commodification and objectification as an interchangeable pump but also a gift of life.

All these uses and definitions of heartbeats make them complex, multiontological, heterogeneous, polysemous, and malleable signs. This is reflected in their nature as central movements that radiate throughout the body via the pulse's blood flow, having a presence in the whole body and then out into the world. The centrality of the heart generates forms of heterogeneity while still being connected. It is a center-periphery relationship in which the center, instead of dissipating, always remains the nucleus.

Throughout this book, I demonstrate how the heartbeat has been a catalyst for complex entanglements between art, technology, corporeality, and ontology that make them polysemous and reveal the heartbeat as a symbolic, aesthetic, and metaphoric device that questions dualities such as terrestrial/extraterrestrial, human/nonhuman, self/other, inside/outside, material/immaterial, sign/index, organism/machine, mechanical/emotional, personal/political, intimate/public, sensitive/rational, vital/morbid, feminine/masculine, life/death. The possibility to rethink our bodies as heartbeat-based interconnects us as humans and with what surrounds us, for we are a constellation of throbbing bodies that inhabit a pulsing Earth, which is part of a pulsating cosmos. This puts into question the dominant Western idea that we are alone in this universe, dominating nature in an autonomous manner. We are part of a constellation of pulsing bodies that depend on other pulsing bodies.

TRACING THE HEARTBEAT

Although today we know the specific physiological factors that make heartbeats occur when the chambers of the heart contract and relax in systole and diastole, the perception of the heart's pulsations has not remained the same over time. It has changed depending on cultural perceptions and according to the advances in science and medicine. In this book, I trace different perceptions of heartbeats and pulses, their measurements and depictions across time. From Chinese medicine to Egyptian and Western medicine, the pulse took different characterizations and forms that led to new ways of perceiving the body.

One of the aims of this book is to trace the aesthetic utilization of heartbeats, in particular how technoscientific advances in monitoring heartbeats and pulses in the nineteenth century—such as René Laennec's stethoscope, Étienne-Jules Marey's sphygmograph and chronophotograph, and Willem Einthoven's electrocardiograph—created a big impact in the development of the arts and to explore how these devices have been historically used in aesthetic ways. For example, the electrocardiograph was used in medical conferences as a spectacle that entertained audiences, and the stethoscope allowed people to hear musical sounds from the heart. These instruments converted the movements of the heart into audible and visual representations and, in turn, transformed humans into technological entities.

The pulse-detecting devices permitted the medical observation of body functions, expanding the interiors of the body to an exterior environment and with it the manipulation of systems that help to record and visualize physiological functions for scientific purposes that then expanded into different artistic fields. Artists would later recognize in the language of these scientific technologies a way of mingling the inner with the outer, the physical with the technological, the emotional with the rational, and data with the flesh.

The use of heartbeats in the arts challenges our conception of the heartbeat and pulse as something intimate and private, as something that occurs under the skin, inside of our bodies in its confined limits. With the development of techniques and technologies, heartbeats are a sign that can be shared, and through that process, artists can communicate personal experiences to their audiences. The heartbeat has become a social phenomenon thanks to the instruments that allow us to hear and see them clearly and show them as data, music, or drawings and in other forms. These instruments created for seeing and hearing from a distance, in hands of artists can become tools of connection and reflection.

Despite the abundance of artistic activity and the centrality of heartbeats as our most vital sign, *Heartbeat Art* is the first historical study of "heartbeat artworks" over the last half-century. This book describes works in detail using archival research, catalogs, interviews, and correspondence; discusses their contexts and development; and examines the larger classes and contours of this neglected area of artistic activity.

HEARTBEATS AND THE ARTS

It is very common to see contemporary artists incorporating heartbeats in their artworks. Though many of them deploy new sensors and technologies, I argue that these contemporary artworks—while unique in their own right—have deep roots in earlier art history. These roots can be traced back to evocations of the heartbeat in ancient dances and music, historical Asian and Western illustrations, in which physicians take the pulse of patients, and to numerous paintings created between the seventeenth and nineteenth century that depict sick, sad, heartbroken female patients with their pulse being measured by a doctor. However, in those depictions, the heartbeat itself was not present but represented, or rather evoked, when a subject touched another person to feel the throbbing of his or her heart, or in poetry, setting its meter to the rhythms of the heartbeat.

The heart has long been a recurrent motif of representation in paintings, drawings, and sculptures—from the multiple depictions of the Sacred Heart of Jesus in baroque *retablos* to innumerous contemporary works, such as Tracey Emin's *I Promise to Love You* (2013) collection of heart-shaped neon lights.[9] But heartbeats and the pulse are signs that reveal the work the heart does rather than depict the heart per se. And while heartbeats have been part of the history of the arts, it was not until the second half of the twentieth century that artists began using the cardiac rhythm as input sourced directly from their bodies.

During the first half of the twentieth century, artists made reference to the heartbeat through mechanical movements or sounds or by mentally following the rhythm of their hearts, as they did not have the necessary technologies to harness the heartbeat directly from their bodies and use it as input. In the 1960s, medical and recording technologies were more accessible instruments for artistic creation. For this reason, although there are mentions of the art of the twentieth century, the focus of this research begins in the second half of the twentieth century.

While innovative in their form and technological infrastructure, contemporary works incorporating heartbeats are full of heterogeneous connotations and intimately related, with a very long history of scientific and artistic experimentation. Hearts, heartbeats, and pulses have been featured in the arts, music, and literature throughout history, but media art history can help us focus on and understand where this technology connects with life, energies, emotions, morbidity, and biopolitics.

A range of technologies has played an active role in the development of artworks using heart rates: speakers, microphones, digital stethoscopes, electrocardiographs, memory recorders, and sensors are some of the tools for creating works that use heartbeats as the main input. This book offers a grounding in this wealth of activity by looking more closely at a number of notable examples that explore some of the connotations of this work and expand on the ways in which artists have incorporated this important vital sign. Also, it explores how medical techniques and technologies have been deployed to record or transform cardiac activity to make it visible and hearable outside the body by means of graphs, light, objects, sound, music, or machines.

I focus on the arts of the second half of the twentieth century to the present, mapping how the technoscience of the nineteenth century had multiple manifestations in the arts flourishing from the 1960s. I discuss in depth various examples of the artistic use of heartbeats, among them

Heinz Mack's cardiographic reliefs that depict two forms of energies: heartbeats and light, Yoko Ono's silent musical compositions that use heart rates as a way to connect people, and Jean Dupuy's installations that show the visual energy of heartbeats using stethoscopes. More contemporary artworks, such as Diana Domingues's digital rituals, merge heartbeats, Amazonian shamanistic elements, and digital technologies. Julieta Aranda's installations uses heartbeats to reflect on economic time and labor. Sasaki's heartbeat drawings use graphs to show that the energy of cities is dependent on people's cardiac rhythms, whereas Christian Boltanski's collection of heartbeats aspires to the impossible: to preserve the memory of all people who have passed away.

The focus on digital artists but also on contemporary artists is intentionally broad. In doing so, it aims to show how these two disciplines that often have been separated—the so-called media arts and the contemporary art—have points of contact that can be clearly reflected in the use of heartbeats, the use of medical instruments, and in the involvement of medicine and doctors in the creation of various pieces.

Hearts, heartbeats, and the pulse have received some scholarly attention in cultural and medicine studies and art history texts, several of which were important for this research. "The Ancient Art of Feeling the Pulse," written by British cardiologist Evan Bedford, is a comprehensive and illustrated review of the history of the practice of measuring the pulse, in which he considers that Chinese and Hindu physicians were probably the first skilled in measuring the pulse.[10] Shigehisa Kuriyama's book *The Expressiveness of the Body and the Divergence of Greek and Chinese Medicine* is a thoughtful and comprehensive account of how medical traditions and conceptions of the body are very different in Asian and Western cultures.[11] Many researchers—such as Jonathan Sterne and Robert Brain—have conducted historical studies highlighting the impact of physiological technologies on culture, aesthetics, and listening techniques.

Victorian Poetry and the Culture of the Heart, by Kirstie Blair, reveals that the heart in the nineteenth century was a subject of general interest. This resulted in a heart-centered arts production in the nineteenth century that engaged the vital organ, as Kristie Blair observes, as "sensationalized, subject to passion and emotion, throbbing and palpitating rather than doing its work unnoticed."[12] In *Matters of the Heart: History, Medicine and Emotion*, Fay Bound Alberti explores the medical, emotional, and cultural significance of the heart from the classic world, in which Galenic theories based on humors played a major role in conceiving the existence of "cold hearts" and charged the heart with symbolisms, to the twentieth-century conception of the heart as a pump and the emergence of modern understandings of medicine that emphasized the role of the nervous system in the perception of emotions.[13]

Some feminist writers have studied biomedicine or the heart from different perspectives, particularly Lynda Birke, whose book *Feminism and the Biological Body* has a chapter on the heart highlighting how science and medicine have influenced our perception of the body and how these disciplines perpetuate gender biases.[14] Anne Pollock, a specialist in social medicine, wrote a paper on the heart and its different approaches as a provocation to think about the heart.[15] Donna Haraway is widely known for her writings in which she talks about technoscience and the fusion of machines and organisms.

Some exhibitions have also explored the heart from diverse perspectives: for instance, *The Heart* (2007), curated by James Peto at the Wellcome Trust in London. Despite the exhibition including a diverse range of objects and notions of the heart from the arts, science, and the every day, both the publication and the exhibition lack an account of artworks using heartbeats but have texts on transplants and the history of the heart that were interesting and helpful.[16]

Some scholarly attention has been given to the hearing and perception of internal body sounds by artists. Douglas Kahn's "Let Me Hear My Body Talk, My Body Talk" examines how military technologies developed during the Cold War were used by male US artists to listen to the sounds inside their bodies in order to create sound compositions or sonic experiments.[17] My research shares with Kahn's paper the intention of seeing how technologies had an impact on the creation of experimental artistic works that explore the stimuli produced within our organisms. However, this study focuses and expands the knowledge on the use of heartbeats in visual, media, and sound arts and includes artists from diverse contexts while also showing how nineteenth-century scientific technologies were the basis of technologies used by artists in the twentieth century to translate heart rates into works of art.

I have paid attention to including creators from different backgrounds and latitudes, as well as to including many cases of women artists who, while being historically overshadowed by their male colleagues, have been continually exploring the internal phenomena of their bodies since the 1960s to the present. My intention is to show that the use of heartbeats is ubiquitous. It is something that unites all individuals, so they are found in the work of artists from all genres and latitudes who use the heartbeat as material in very different and rich ways.

Heartbeat Art falls mainly within the history of the arts and, through its emphasis on technoscience and the arts, the history of media would be the most hospitable area of discourse. The history of media is defined by a combination of skills, disciplines, and sources for studying mostly digital artworks or art-science collaborative art projects.[18] As such, media art history is an interdisciplinary method for revealing and understanding the evolutionary history of media and for understanding the developing relationships between technologies and the arts.

I draw on histories of the arts of the 1960s and 1970s, the history of music and sounds in the arts, contemporary art history, intellectual history, cultural studies and medical literature, and investigations of technology and scientific instrumentation in order to understand heartbeat-related artworks as being deeply connected to their immediate contexts, while also being related to longer trajectories of technoscientific and artistic experimentation. The historical study is based on archival research from primary sources and on wide-reaching published and online historical and contemporary materials, including books, catalogs, art magazines, journals, and artists' documents from the 1960s to the present.

As this is a topic that has not previously been mapped, the study also draws on interviews and correspondence from artists who have used heartbeats in their work as primary sources of information. Among those I interviewed or exchanged emails with are Teresa Burga, Diana Domingues, Heinz Mack, Linda Montano, Brian O'Doherty, Éliane Radigue, Catherine Richards, Sasaki, Julieta Aranda, and Rafael Lozano-Hemmer. Another major contribution to this book is the connecting of this new primary material with a wide range of secondary sources, mostly articles in which these artists were talking about heartbeats. This study represents the first time that this material has been brought together and constitutes an in-depth analysis of the various statements, histories, and heartbeat works of all the artists addressed in this book.

As a consequence of the complexity and variety of connotations of the heartbeat works, this book uses an interdisciplinary method that analyzes oeuvres through different disciplines in order to produce new and expanded ways of understanding material culture. The different conceptualizations of these elements are analyzed and entwined, therefore mapping possible new narratives of heartbeats in the arts. This web of relations demonstrates the relation of art, developing a variety of nonlinear forms and interpretations, which reveal that art and artifacts have had an intimate relation in the creation of aesthetic experiences related to the body.

CARDIOGRAM OF THE BOOK

This book is structured in seven beats or chapters, each of which explores different aspects of heartbeats in relation to the arts. Since each beat has an introduction, I will forgo repeating them. The first beat, "Measuring," serves as a setting ground on how historical advances in pulse measuring and scientific instrumentation have an impact on the sonic and visual representations of the heart by artists in the nineteenth century and the early twentieth century. I also show how there was a transition in medicine from feeling pulses to listening and observing pulses, and how the medicalization of the female body played an important role in the development of medical technologies. This section sets the stage for the next sections in which there are in-depth studies of various contemporary artists.

The rest of the beats are focused on the arts from the 1960s to the 2000s. The chapters—"Space," "Electrical," "Time," "Transplants," "Surveillance," and "Regenerations"—are based on some capacities of the heart, as a sign that occurs in a specific space but expands beyond corporeal and terrestrial limits, as a powerful source of electromagnetic energy in the human body, as a sign that is associated with the identity of a person, as an organ that has the faculty of regenerating in physical and emotional ways, and as a source of connection between different people. These chapters start in the cosmos, where we can find "heartbeat" signals of stars and astronauts, and the last beat ends with a reflection on the finiteness of human life. Although the categories of the beats are set, many of the works described could go in other beats, mirroring the contracting and expanding character of the heart as a sign that is generated in a specific area but permeates the whole body.

I also show how medical instruments developed by scientists, who usually focus on narrowing down the meaning and uses of the object of observation with a logical and rational focus, have been used by artists in various ways that expand the meanings, applications, and functions of our heartbeats and pulses. Heartbeats and pulses are not just bodily functions, dictionary definitions, or scientific descriptions of a phenomenon; they are the imaginary uses that we ascribe to them.

While my journey into understanding the role of heartbeats in the arts began a decade ago, this book introduces a whole new constellation of uses of heart rhythms. As the Mexican playwriter Emilio Carballido said, knowledge is there, always beating, waiting to be found. But more importantly, there is no knowledge without the heart because the heart is knowledge. I hope that with this flow of information that crosses time, expands from the interior to the outside of the body, and circulates the earth, some unexpected channels and currents will be revealed, helping us understand and perceive that constant ticking that accompanies us faithfully day by day, and which is charged with so many biological, symbolical, emotional, and political connotations.

1

MEASURING

Today, you can sense your pulse in your wrist, hear your heartbeat with a stethoscope, and see your heart's wavy patterns on an electrocardiogram. This extension of the heartbeat into another space was made possible by a long-standing tradition of pulse studies from various civilizations that culminated in the nineteenth century's fascination with the heart. This interest was reflected in important developments in the study of the structure and function of the heart, which in turn gave birth to new technologies, such as the stethoscope, the sphygmograph, and the electrocardiograph. These instruments opened a great universe of heart sounds, graphs, and electromagnetic waves that were amplified and made an image of the heartbeat. The creation of various cardiac technologies marked the transition from hand feeling to instrument tracing, and therefore, from closeness to distance. However, the heartbeat always remained more than its measurements, as it contained metaphors and affections.

This chapter traces the development of the art of listening to, observing, and characterizing the pulse using such apparatuses, which helped produce clearer explanations of how the heart functioned and led to new perceptions of the organ as a pumping and essential machine. The histories of these techniques and medical instruments were captured in many paintings of the time, and the hearing of the sound and the visualization of the beating heart had an aesthetical character. Similarly, the waves created by the heartbeat were turned into graphs and images that had their own particular form and language. Presentations of these instruments featured in scientific-artistic entertainment, through which heartbeats and the arts became intimately connected. The heart sounds and graphs produced by these apparatuses had a considerable impact on modernist aesthetics, influencing works by surrealists, futurists, and other modernist currents. This will have an impact on the production of the arts of the twentieth and twenty-first centuries through works that directly connect the arts and these technologies.

FEELING PULSES: INDIAN, CHINESE, EGYPTIAN, GREEK MEDICINE

In a small watercolor painting, you can see *An Ayurvedic Practitioner Taking the Pulse* (c. 1830; figure 1.1) of a patient on his wrist in India, where one of the oldest registers of pulse palpation is found in Ayurvedic medicine in the texts of the Vedas. In the "Science of Sphygmica" by Indian physician and philosopher Kanada (600 BCE), a variety of pulses are described, which depend on the three main humors of the human body: bayu/vata (air), pitta (bile), and kaph/kapha (phlegm). The practice of measuring the pulse is not new. The thin layer of tissue that covers our body has consistently functioned as a natural interface between the inside and the outside. Three basic methods for feeling and measuring the pulse have been used since the earliest times: inspection, palpation, and auscultation.

The sensing of the pulse on the wrist has also been practiced since ancient times in traditional Chinese medicine. The Yellow Emperor's Cannon or *Neijing* (200 BCE) and Wang Shuhe's *Maijing* (300 BCE) are classical writings that register the early Chinese practice of measuring a patient's *mo* or pulse through a sophisticated system that emphasized a natural and holistic approach to illness. As shown in Zhou Pei Qun's *A Mandarin Doctor Consulting a Patient* (ca. 1800; figure 1.2), the practice was performed in a quiet environment after the patient relaxed and placed an arm on a cushion. The doctor palpated each wrist with light, moderate, and firm pressures, locating the index, middle, and ring fingers on the radial artery. Chinese physicians felt the flow of *qi* (a vital force) and the balance of yin and yang through the pulse, focusing on the regularities and irregularities of each pulsation. They also took into consideration the

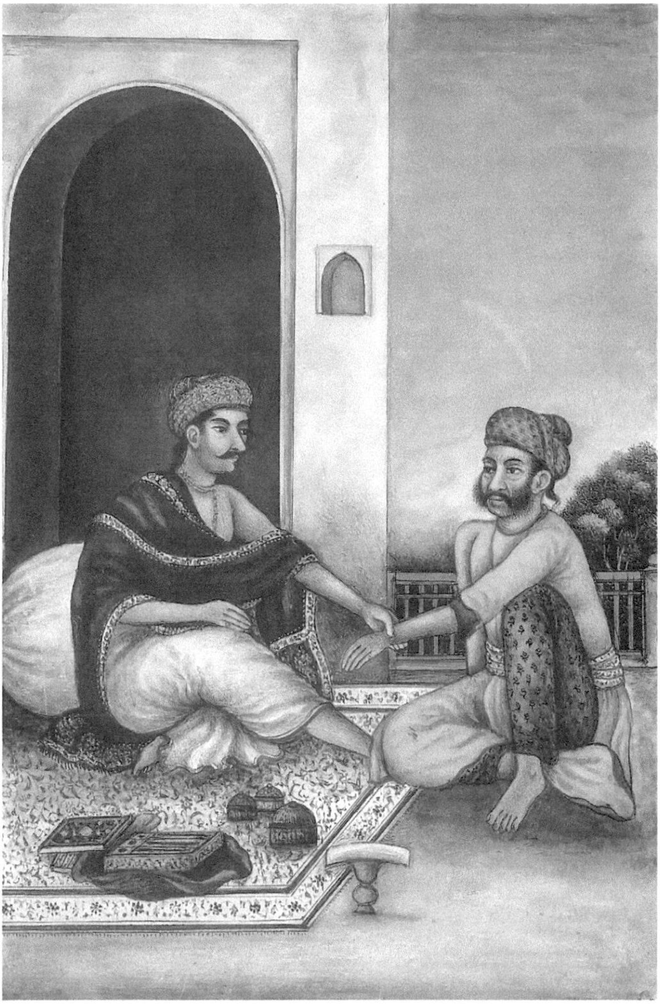

1.1

Unknown Delhi artist, *An Ayurvedic Practitioner Taking the Pulse*, c. 1830. Wellcome Collection.

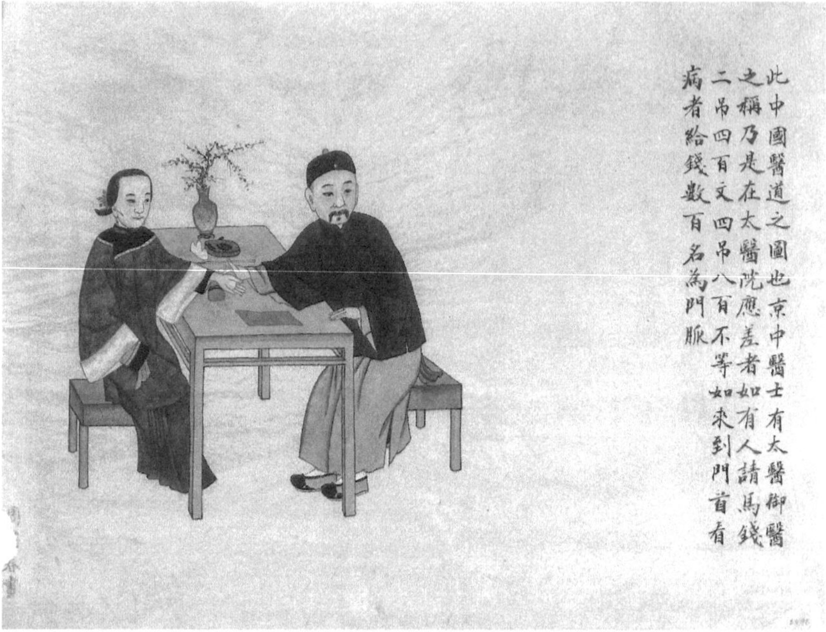

breathing, temperature, and emotional state of the person. By the palpation method, and without the necessity of interrogating patients and knowing their symptoms, doctors were able to know through their fingers not only the condition of the heart but of all other viscera.

Ancient Chinese medical literature identifies around twenty-four different types of pulses or images: "soft," "deep," "weak," "steady," etc.[1] Each organ also had its particular pulse and a predominating season in a calendar. Their combination indicated states of the body; for example, a slippery and rapid pulse may be a sign of gestation. The seasonal aspect of each pulse was a reflection of the macrocosms in the microcosmos: a

1.2

Zhou Pei Qun, *A Mandarin Doctor Consulting a Patient*, ca. 1800. Wellcome Collection.

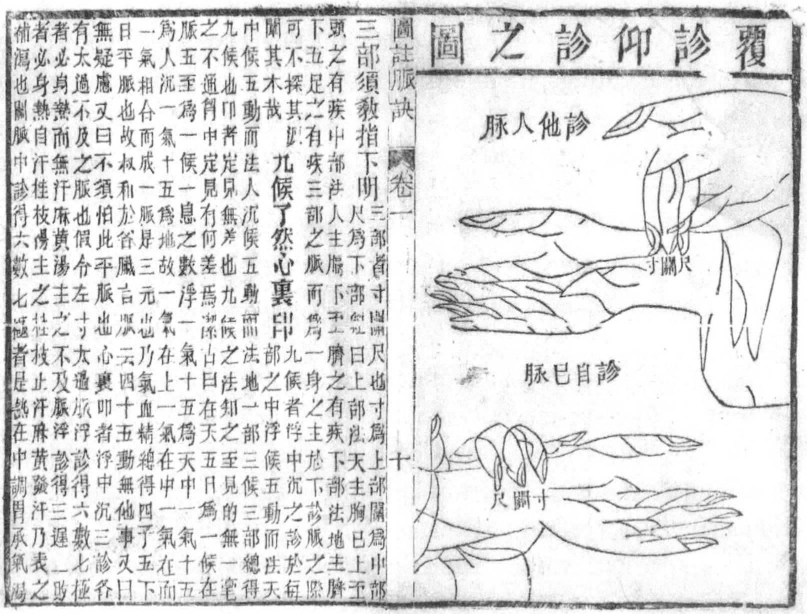

misplaced pulse or a seasonal variation signed an alteration in the order, meaning an illness (figure 1.3). Characters and poetic descriptions rich in metaphors and analogy were used to describe the pulses. "Floating and extremely forceless like a silk thread in water" or "like wind blowing hair or scattered leaves" are some examples of their imaginative, intuitive, and nature-based descriptions.[2] The art of palpation, with its marvelous forms of knowing the hidden secrets of the body, also extended to other Asiatic regions, such as Japan and Tibet.

Dating as early as 3000 BCE, Ancient Egyptian Medicine also practiced pulse examination using the haptic method and distinguished the

1.3

Ancient Chinese document showing the measuring of the pulse. Wellcome Collection.

different essential components of the cardiovascular system: the heart, the blood vessel, and the pulse. The Papyrus Ebers (1550 BCE), one of the oldest medical writings, mentions the use of pulse palpation:

> There are canals (or vessels) in it (the heart) to every limb. Now if the priests of Sekhmet or any physician put his hands (or) his fingers on the head, on the hands, or his fingers on the back of the head, upon the two hands, upon the pulse, upon the two feet, he measures the heart, because its vessels are in the back of the head and in the pulse; because its pulsation is in every vessel of every member.[3]

Greek medicine is based on Egyptian medicine, and although the Greeks were looking for rational explanations, seeking a distancing from Egyptian religious perspectives, they were also practicing a vitalist notion of medicine. The haptic method was known and dominant for many centuries in ancient Greece. Hippocratic medicine recognized the valuable use of arterial pulse as a symptom and was used for detecting and measuring the grade of diseases. The pulse became considerably more important for the School of Alexandria. In the fourth century BCE, Praxagoras of Kos (340 BCE) was among the first Greek physicians to examine and write about the pulse. He defined the pulse as a movement of the arteries distinct from morbid tremors. Like his predecessors, he attributed the origin of the pulse to *pneuma*—a "vital air" from which everything originated, which traveled along arterial layers while veins contained the blood.[4] For Praxagoras, the pulse was an important diagnostic tool, helpful for tracing diseases, which could manifest through different kinds of pulsations.[5]

A few years later, Herophilus of Alexandria (335–280 BCE) was the first to consider that pulsation was entirely involuntary and was caused by the contraction and dilatation of the arteries in accordance with an impulse received from the heart. His book *On Pulse* was a foundational text for ancient Greece pulse theories, which declared that the arteries were multifunctional, transporting *pneuma* but also blood.[6] Before any other physician, he distinguished between veins and arteries, and pulsations (*sphygmos*) from palpitations (*palmos*),[7] and measured the pulse rate using an aesthetic and innovative apparatus: a *clepsydra* or portable water clock that contained a specific amount of water and was calibrated according to the age of the patient.[8] Erasistratus (304–250 BCE), another physician from the School of Alexandria, stated that the heart was the leading organ of the circulation system: the heart contracted, and the arteries dilated as a passive expansion of the vessel.[9] Erasistratus thought that *pneuma* was the cause of the heartbeat and the origin of the natural heat of the body, as well as being responsible for the body's nutrition processes.[10]

On the heels of Herophilus's work, the prominent Greek physician and philosopher Galen of Pergamon (129–ca. 216 CE) added to the understanding of the heart and arterial pulse in his seven treatises on the pulse.[11] Through observation, experimentation, and palpation, Galen studied the different rhythms of the pulse based on the degree of dilation of the arteries—strong, medium, or weak—which were detected through the skin:

> We detect in several parts of the skin some sort of motion, and this not only by pressing down on them, but sometimes with our eyes as well. Moreover this motion is found among all healthy people in many parts of the body—of which one is the wrist. [In such places] we can clearly detect something coming from below up to the skin and striking against us; after the beat, sometimes it distinctly moves away and then pauses, and sometimes right after the beginning [of motion] it appears to pause, and then comes again and beats, and then it goes away again and rests. And this process goes on in the entire body, from the day we are born until we die. It is this type of motion that all people call the pulse.[12]

Galen thought of the heart as an enclosed structure that, by its "innate heat," dilated when drawing blood from the liver into its right chamber, converting it into a more refined type of blood. This arterial blood traveled through pores to nourish the lungs, continuing its circulation into the left heart chamber, which was a *pneuma* pump that sent the vital substance into the arteries, producing the pulse.[13] Galen incorrectly stated that diastole—the period of relaxation of the heart muscle when the chambers are filled with blood—and not systole—the period of contraction of the ventricles of the heart—was the true movement of the heart but correctly found different kinds of pulses based on the degree of arterial dilation, which were usable for diagnosis.

The Persian polymath Abu Ali Ibn Sina, known as Avicenna (981–1037 CE), wrote the *Canon of Medicine*, a medical treatise regarded as a "medicine bible" that remained important in the East and West until the sixteenth century. The pulse's significant presence in the manuscript is announced on the cover of some editions in which a *Doctor Taking a Woman's Pulse* appears (figure 1.4). In the text, among many contributions, Avicenna correctly conceived pulse for the first time as a wave rather than a straight impact and extended Galen's studies about arterial pulse rhythms. Avicenna described for the first time in detail some cardiac diseases, and considered both physiological and psychological factors, and diet and drugs to treat patients.[14]

During the Renaissance, the human body intersected science and art, and artists studied anatomy to understand the forms of the body. The multifaceted practice of Leonardo da Vinci (1452–1519) would be the greatest example of this crossroads in practices. Leonardo was a polymath who worked across disciplines, including music, painting, and science. He wrote more than fifty pages between ca. 1508 and 1513, many dedicated to cardiac studies. His interest in dissection—which he mainly practiced with animals—led him to correctly describe the heart as a four-chambered structure. He also recognized the heart as the center of circulation, though this was clearly in conflict with the theories of Galen, which considered the liver as the originator of the vascular structure.[15] Nevertheless, Leonardo never fully understood how the heart pumped or the circulatory system functioned.

Galen's understanding of the pulse and medicine dominated clinical practice for many centuries until the English physician William Harvey (1578–1657), with his groundbreaking studies, which received strong criticism at the time, correctly attributed the generation of the arterial pulse to the contraction of the left ventricle and located the source of heartbeats in the right atrium of the heart. Harvey elucidated that it was in systole and not diastole that the heart drew blood into its chambers and conducted the blood out into the arteries, producing the arterial pulse.[16] He also accurately described the blood circulation system of the human body. However, Harvey did not devise any specific instrument for assessing and visualizing the frequency of the arterial pulse; his studies were instead based on tactile methods. Feeling the pulse on the wrist was already a long standard practice, and it served as the conventional representation of the medical profession for many centuries. Palpation remained an important

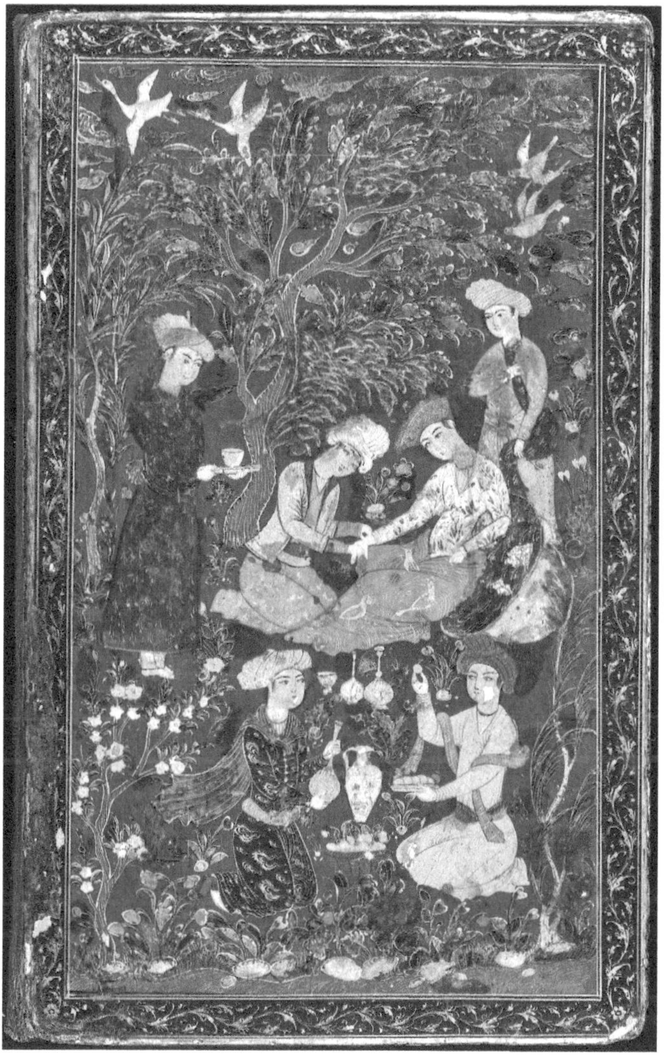

1.4

Doctor Taking a Woman's Pulse. Persian lacquered binding board cover of Avicenna's *Canon of Medicine* manuscript, 1631. Wellcome Collection.

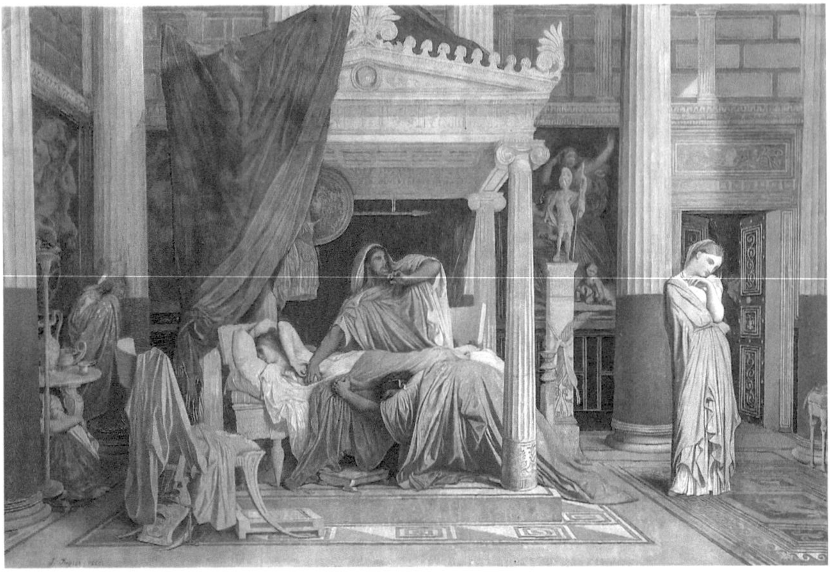

method throughout centuries for knowing what was happening inside of the body. The image of touching the wrist represented the pulse in art for many centuries, and palpation was considered a method for revealing not only ailments but also affections and afflictions. This was especially evident in a Greco-Roman legacy illness that combined medicine and emotions: lovesickness.

Jean-Auguste-Dominique Ingres's *Antiochus and Stratonice* (1866; figure 1.5) depicts the ailment by painting a pale lad miserably lying on his bed while Erasistratus is measuring his pulse rate. The doctor detects that Antiochus's heart agitates in the presence of Stratonice, his father's wife. In this love triangle drama, the central figure is lovesickness. The notion of the

1.5

Jean-Auguste-Dominique Ingres, *Antiochus and Stratonice*, 1866, from Patricia Condon, *Ingres, in Pursuit of Perfection* (Louisville, KY: J. B. Speed Art Museum, 1983). Image © Fabre Museum of Montpellier Méditerranée Métropole.

existence of a pulse peculiar to love, initiated by Erasistratus, was extended by Galen, who influenced the long conception of lovesickness as a female pulse-detected physiological affliction and humoral imbalance that was strong enough to become lethal or to turn people insane.[17]

Various paintings of the seventeenth century, all created by men who also led the artistic field, show how the pulse of lovesickness supposedly affected only females. *The Physician's Visit* (c. 1660; figure 1.6) is one of the over eighteen paintings by Dutch artist Jan Steen depicting quacks checking the pulses of pale, sick females and holding flasks of urine to determine pregnancy and erotic melancholy. The abundance of paintings and narrations representing females as victims of lovesickness, not men, responded to a generalized perception of women as more emotionally vulnerable than their male counterparts, echoing the distinction Aristotle drew between "hot hearts" and "cold brains." This gendered assumption that still stands today permeated all cultural forms, including the sciences and the arts, during the nineteenth century.

In Enrique Simonet's *And She Had a Heart!* (*Anatomy of the Heart*) (1906; figure 1.7) a beautiful young woman is lying dead on a hospital table, while a doctor observes her extirped heart. Her heart is gendered as a container of feelings and a sample of anatomical study. Over many centuries, women were objects under the scrutiny of the male-lead medical circle, which invented a series of female-only diseases.[18] In the pulse was found more than blood passing, it was the language of the heart that passed throughout the body, and what was measured and controlled by doctors was not only a physical function, but also sexual features and functions.

As noted by Michel Foucault, sex was the focus of biopolitics in the nineteenth century, with "indeterminate medical or psychological examinations, to an entire micro-power concentrated on the body."[19] The pathologizing of the female body and the idea of quickness of heartbeats as a psychic energy and a form of lust continued in Sigmund Freud and his numerous studies on psychopathologies, such as melancholy and hysteria. Hubert-Denis Etcheverry's *Une Consultation* (1900) depicts Freud measuring the pulse of a weak young woman with one hand on her wrist and the other holding a clock as he tries to detect hysteria by one of its various symptoms, "the involuntary retraction and trembling of the wrist when you take hold of it to fell the pulse."[20]

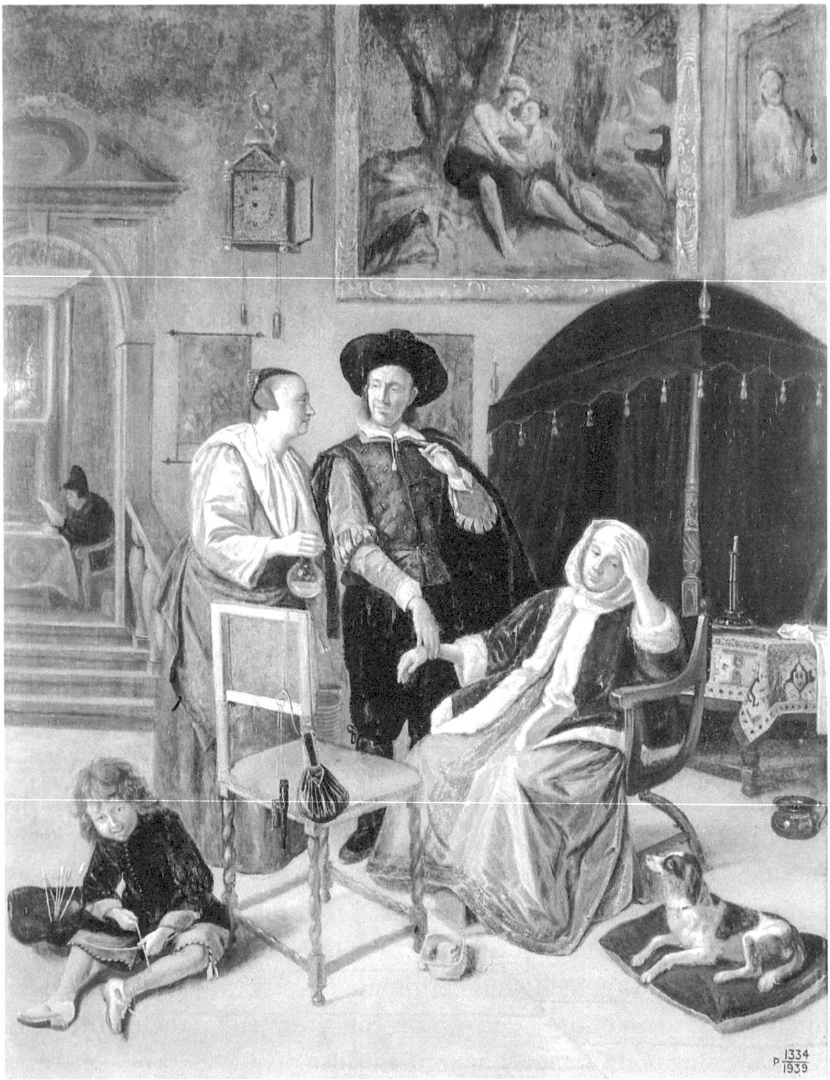

1.6

Jan Steen, *The Physician's Visit*, ca. 1660, Oil on wood, 51×41 cm. Wellcome Collection.

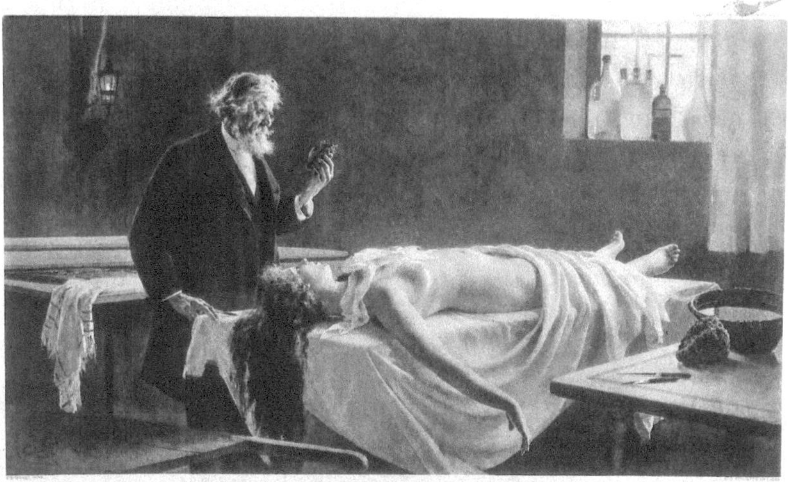

The use of the clock was the first attempt at creating an "objective" mediator between our vital signs and subjective perceptions. Its use can be traced back to Herophilus's clepsydra and Galileo (1564–1642) and Sanctorio Sanctorious (1561–1636 CE), who used a pendulum for measuring the pulse. But John Floyer (1649–1734 CE) started the use of watches to calculate the number of beats per minute. He made several observations on the pulse in his treatise *The Pulse-Watch* (1701): "I counted the pulse after a great passion of anger, and it beat 104. . . . I found the morning pulse to exceed near 10 beats, if I drank ale or wine at night; Fasting and being very cold, reduce my pulse to 62."[21] However, the practice of using a watch for timing the pulse rate was popularized by the Irish physician Robert

1.7

Engraving "Der Anatom" after Enrique Simonet's *And She Had a Heart! (Anatomy of the Heart)*, 1890. Photogravure by R. Schuster, 1907. Wellcome Collection.

James Graves (1796–1853 CE), who established as a key part of the practice of examination the usage of a clock for the count of pulses.[22]

Heartbeats became a gendered mechanism for revealing feelings, and their measuring techniques were ingrained in a heteronormative scientific rationale. In fact, as we will see below, it was the fear of the female body and women's emotions that gave rise to instruments for measuring at a distance, without touching, the talking of the heart.

LISTENING HEARTBEATS: RENÉ LAENNEC'S STETHOSCOPE

Picture a nineteenth-century medical ward in which one of the beds contains a sick woman with her back bent forward. A physician, in front of a dozen of his alumni, places his ear on the back of the patient to hear her body sounds. The auscultation scene that you are seeing is *A Hospital Ward during the Chief Physician's Round* (1889; figure 1.8) by the Sevillian painter Luis Jiménez Aranda (1845–1928). In the composition, the female is the sick and frail object, while men are scientific beings learning that direct hearing of corporeal sounds, much like pulse-taking on the wrist, reveals what is happening inside the body.

The practice of immediate auscultation was commonly taught and practiced in Europe during the nineteenth century, but its roots go back much further in time. Physicians since Hippocratic times have practiced immediate auscultation by placing their ears on the back and chest of a patient, looking to detect inner noises that could be a sign of disease.[23] Proof of this activity lies in a passage from Hippocrates's treatise, *De Mobis*, in which he declared that "if in applying the ear during a certain time on the side, you perceive a noise like that of boiling vinegar."[24] In spite of its effectiveness, the auscultation method remained neglected until its re-emergence in 1761 when the Viennese physician Leopold Auenbrugger (1722–1809) started studying the intero-ambience of the body, tapping on the torsos of his patients like a percussionist in order to detect diseases via the "fullness or emptiness" of their body sounds.[25]

In his monograph, *Inventum Novum ex Percussion Thoracis Humani Interni Pectoris Morvos Detengendi* (1761), Auenbrugger described the gamut of body noises as drumlike: "The sound the thorax gives is observed to be

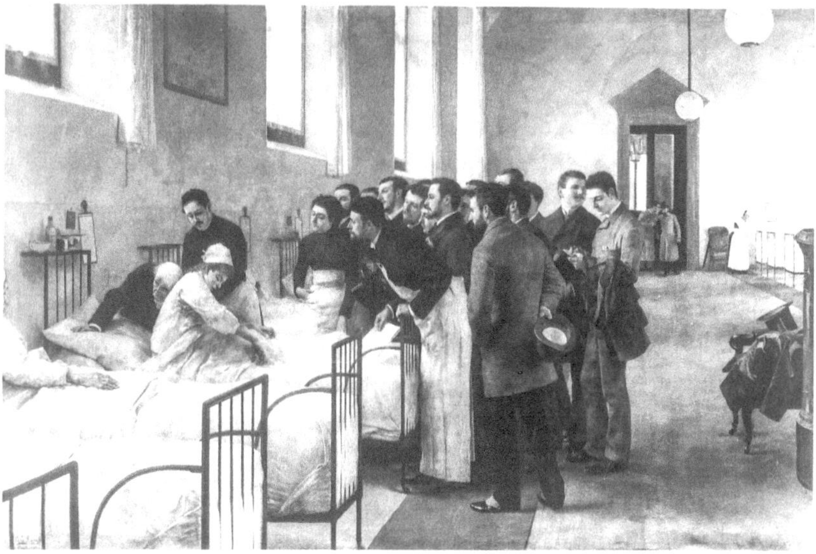

such as in drums [tympanis] when they are covered by a cloth or by another fabric made of thick wool."[26] Clear sounds were considered normal, whereas heavier or duller noises were signs of abnormalities. Although Auenbrugger's connection of drums and medicine was ingenious, the relation between heartbeats, drums, and medicine has a longer thread in history.

Since ancient times, indigenous cultures from Africa, Asia, and the Americas believed that heart and drum beats were interconnected and affected one another. For millennia, African civilizations have regarded ancient drums, like the *djembe*, as potent healing instruments with their

1.8

"Charcot at the Salpêtrière," after Luis Jiménez Aranda's *A Hospital Ward during the Chief Physician's Round*, 1889. Wellcome Collection.

own language and the ability to influence human health and behavior.[27] The Aztecs also believed that the sounds of drums, such as the *huéhuetl*, represented the universe's heartbeat and could thus be used to heal both body and spirit.[28] Similarly, for centuries, ancient Korean and Japanese female shamans have used drumming to drive away malign spirits that could enter the body and cause illness.[29] The priestesses from Egypt and also in Rome are reported to have used the drum to honor their deities and connect to the healing boundless energy of life.[30] One of the passages of the Papyrus Ebers, which reads, "Examine his belly and knock on the finger," demonstrates that direct percussion of the body as an inspection method has been done since ancient times.[31]

While Auenbrugger did not invent the method of using drumming for healing, he did continue the historical legacy of body drumming by writing a treaty that described it as a scientific diagnostic tool. The use of percussion for exploring the body's soundscape represented a change in the medical diagnostic method. Instead of understanding diseases as the product of a humoral imbalance of bodily fluids (blood, black bile, yellow bile, and phlegm), they were seen as maladies whose signs could be discerned through a body-noise-based language. In this manner, the senses of the doctor became the key to detecting a disease, thus relegating to a secondary role the narrations of a patient and their feelings, which until then had been the main basis for diagnosis.[32]

Auenbrugger's percussion method gained considerable attention across Europe years later when Jean Nicolas Corvisant (1755–1821), Napoleon's physician, found body-tapping valuable for diagnosing diseases of the chest. Until then, Auenbrugger's observations were considered nonscientific, derived from his amateur interest in music and from his observations of his father tapping wine barrels in his cellar.[33] Another factor that slowed the acceptance of body percussion was that it required skin contact, which was regarded as indecent, and the full exposure of the patient's body to avoid extraneous noises.[34]

The risqué character of direct auscultation is clearly alluded to in Jules-Abel Faivre's *The Examination* (1898; figure 1.9). Here a young and flushed woman lying on a bed lifts her white robe to reveal a voluptuous body for examination by an elderly doctor. He is pressing his ear toward his patient's breasts to listen to her chest sounds, but while he seems

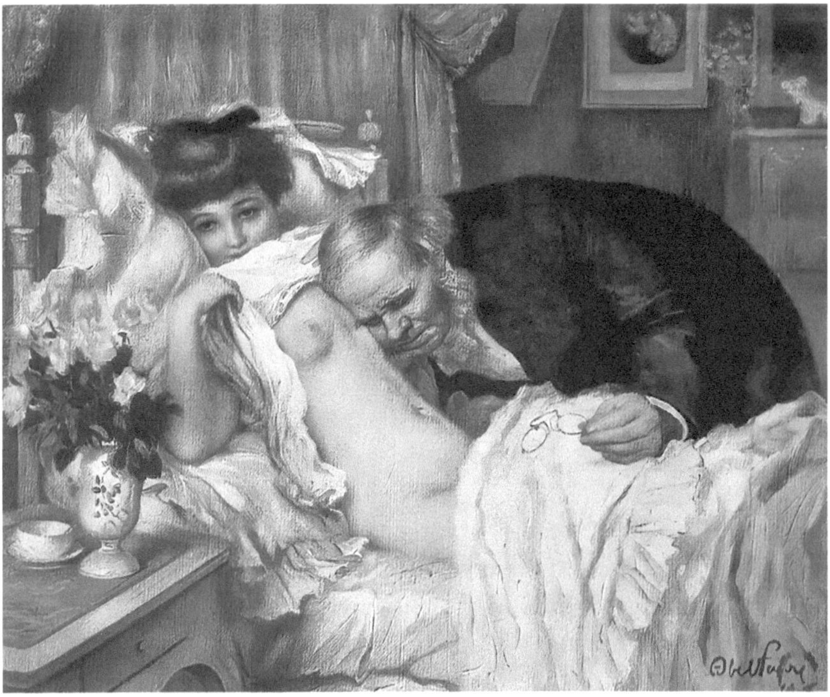

1.9

Jules-Abel Faivre, *The Examination*, ca. 1898. Chromolithograph after A. Faivre. Wellcome Collection.

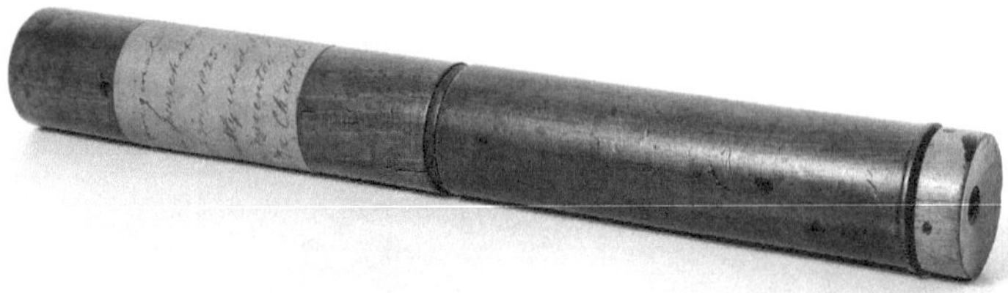

focused on hearing the lady's internal corporeal noises, her coy expression and her exposed body demonstrate why direct auscultation of women contravened the standards of morality and respectability of that time. This painting is a testimony of how the body in the male-led medical field is not seen as a neutral organism but as a site where gender, appearance, race, and class play a determining role, and in which the female body and the perception of its pulse has been seen as something sexualized, before which it is necessary to set boundaries.

Seeking to avoid improper closeness to the female body and also the smell and fatness of some of his patients, Corvisant's former pupil, René-Théophile-Hyacinthe Laennec (1781–1826), created the stethoscope (figure 1.10) in 1816. This small instrument transformed medical diagnosis at the same time as it increased a culture of distance. The French physician's invention was suggested to him during a medical consultation with a young and plump woman. He considered it indecent and inefficient

1.10

Monaural stethoscope; Laennec type. Wellcome Collection.

to practice direct auscultation, so he devised another solution. Inspired by an imaginative children's game, in which a long and hollow stick is used to amplify the tiny sounds of scratching, Laennec rolled a piece of paper to create a sort of trumpet, which he placed as a bridge between his ear and the patient's breast. He was delighted and surprised that by this simple method, he was able to hear clearly, as never before, the beating of the heart. He realized that this technique could be extended to hearing and recognizing many other sounds in the thoracic viscera.[35]

After this discovery, Laennec built several models of his instrument and experimented with various materials. After discarding metal and glass, as they do not convey sound well, he built the first prototype of his device with lightwood, a material that clearly transmitted the array of sounds of the heart.[36] Laennec called his device the stethoscope—a term that, in Greek, literally means looking at or into the chest. Elegantly held "like a writing-pen," the instrument was an extension of the ear trumpet that modified the aesthetics of touching the body of a patient directly while permitting the clear audition of the invisible inner body world.

As a result of this semiaccidental discovery and with a very simple device, Laennec revolutionized medical diagnosis. He termed his technology *mediate auscultation*: the hearing of the body with the aid of an instrument. This method transformed the epistemology of medicine by ushering in a new emphasis on physical distance and technologized methods. It changed the doctor-patient relationship by reinforcing a separation not only between the examiner and the examined but also between classes and genders.[37] This happened at a time when greater social mobility was transforming Europe with the rise of the Industrial Revolution, alongside rapid urbanization and new systems of education and health.[38] Furthermore, as Jonathan Sterne has noted, the stethoscope and mediate auscultation was "a new approach to reason and the senses, in which listening moved away from the ideal of spoken exchanges between doctor and patient into the quiet, rhythmic, sonorous clarity of rationality."[39] And with this, speech and touch, two modalities of communication, were supplanted by silence, distance, and indirect listening.

Although Laennec's systematized method, which linked sound rhythms to the pulse for the detection of certain diseases, gave a relevant role to the sonic and musical aspects of the body, the relationship of rhythm, music, and pulse goes back to antiquity. Indigenous cultures from Africa, Asia, the Americas, and the Antipodes have associated drums with music and the heartbeat of Mother Earth. Drumming and dancing in Africa were regarded as an emotional vehicle to connect to the heart of the universe and the heartbeat of the community.[40] For millennia, Aboriginal Australians have had a rich medical, musical, and dance heritage, frequently using clapping sticks and didgeridoos as vibrating healing tools with a natural sound that was in tune with the land's heartbeat. In Māori culture, hakas are healing collective and connective dances in which each stamp of the foot on the floor represents Papatūānuku's heartbeat. Drumbeats were associated with the heartbeat of the people, animals, and Mother Earth by First Nations in Canada, who saw drum-based music as a means of communicating with other living beings and deities and as a form of healing.[41]

In the West, Herophilus was among the first thinkers to compare pulsation with an art form: music. For many physicians of the time, music was a heuristic device, a guide enabling analogies: Herophilus thought that the pulse had upbeat and downbeat units that could connect metrical-musical rhythms with pulse rhythms, including the rhyme and meter of poetry.[42] The pulse's rhythm, intensity, and speed were also relevant for Galen, who distinguished more than twenty different kinds of pulses based on rhythmic sensibility and musical ratios. Pulses were "strong," "medium," or "weak" and had three different dimensions: length, breadth, and depth.

For the Persian polymath Avicenna, pulse was also intimately related to music; for him, the true nature of music was to be found in pulse, and pulse beats were comparable to rhythmic beats in both their qualities and harmonic modes. Like Galen, Avicenna thought that only a person familiar with the science of music could grasp these relationships.[43] The discoveries of Galen and Avicenna strengthened the relation of pulse with music and had a considerable impact not only on physiologists but also on music creation during the Middle Ages and the Renaissance. It was believed that music was inherently linked to rhythms and things that could not be seen. The throbbing of the pulse, for example, was conceived as a fundamental, inherent musical harmony in the body and soul of the human being.[44] The music of the spheres was determined in Boethius's *musica humana* by the rhythms of the human pulse but also by hormonal cycles and the many identifiable patterns of breath and heartbeats. All these unseen physiological functions were attuned to the universe in melodious and proportional ways.

During the fourteenth and fifteenth centuries, many physicians extended the relation of pulse with musical proportions. The Italian doctor Gentile da Foligno (d. 1348) considered that the intensity of a heartbeat corresponded to musical measurements in which strength and weakness related to high and low pitches and fast and slow tempos to motion and rest, respectively.[45] The Spanish theorist Bartolomeo Ramos de Pareja (ca. 1440–1491) considered that the human pulse was the measure of musical tempo. This is an idea shared by Leonardo da Vinci, who discussed the temporal nature of music using the term *tempo armonico* for a beat of music in his famous *Paragone*, a comparison of poetry, music, and painting. Leonardo considered "the interval between two beats of the pulse was half of a musical tempo."[46]

Following the notion that there was a correlation between the beats of music and the heart, the Polish physician Josephus Struthius (1510–1569) attempted the first graphic representation of the pulse using musical notes for each rhythm. In the following century, a musical interpretation of the pulse was extended in Robert Fludd's *Pulsus* (1631), Samuel Hafenreffer's *Monochordon Symbolico-Biomanticum* (1640), Athanasius Kircher's *Musurgia Universalis* (1650), and Francoise Nicolas Marquet (1687–1759).[47] The latter proposed the understanding of the "natural regular pulse," specifically the spacing between heartbeats, as similar to that of a slow, graceful dance: a minuet.[48]

While the practice of relating music with pulse thus had a long tradition, Laennec took it a step further by creating an apparatus that allowed physicians to hear the sonic universe of the heart and the lungs more accurately. With his innovative device, Laennec was able to identify, analyze, and classify hundreds of sounds, connecting them to anatomical diseases and lesions (figure 1.11). His most substantial contributions were to maladies of the lungs, such as tuberculosis, but he also contributed to the study of the heart, describing a variety of noises and murmurs. Before the nineteenth century, the body was commonly considered to be largely silent, only occasionally making audible sounds, but, as Sterne observes, with indirect forms of auscultation, "speaking patients with mute bodies gave way to speaking patients with sounding bodies."[49] As the *bruits*

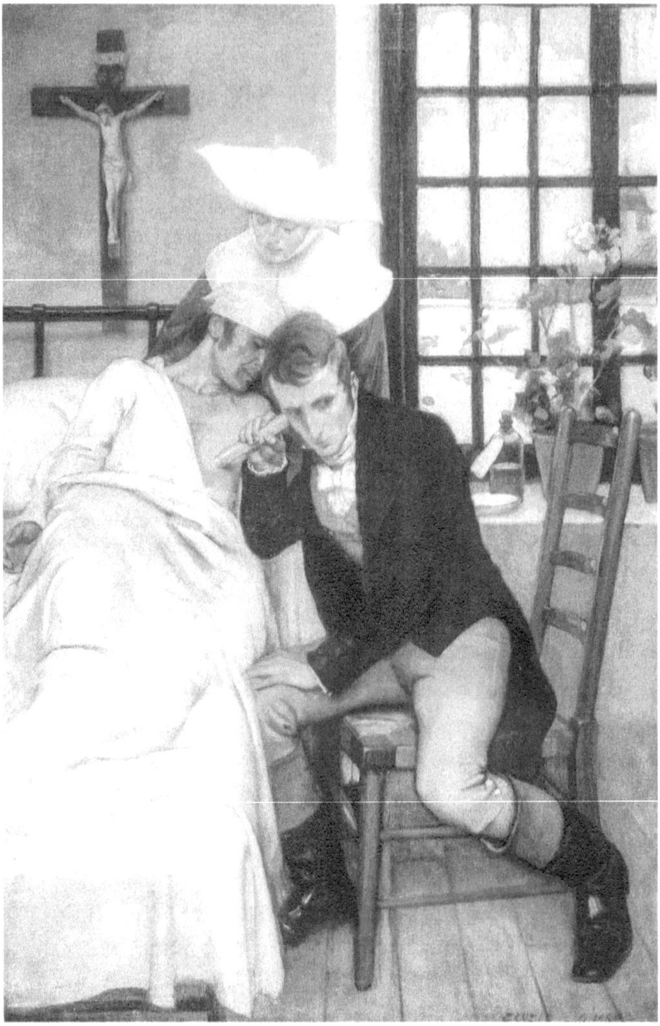

1.11

Ernest Board, *Laennec Listening to the Chest of a Patient*, ca. 1910, from Alan E. H. Emery and Marcia L. H. Emery, *Medicine and Art* (London: The Royal Society of Medicine Press, 2003).

that Laennec found in his patients' chests made a pleasing melody, he described them using musical and poetic terminology.

The spectrum of tuneful noises that Laennec found in the heart was wide. Healthy hearts emitted two sounds: one *bruit*, clear and swift; another, slower and duller.[50] When diseased, he found the heart orchestrated multiple tonalities. Some vibrations in the thorax were like "a fine harp-string slightly rubbed with the end of the finger."[51] Other artery sounds were evocative of "military music, on a march, every now and then interrupted by the hoarse roll of the drum."[52] For Laennec, cardiac activity composed "a certain succession of musical tones" with rhythms that altered between action and repose.[53] Thus, he conceived of the thorax as a music box, with complex compositions deftly executed by the organs. What Laennec was hearing was music coming directly from the heart. He even scored the "songs" coming from ailing bodies:

After assuring myself that the sound came from the artery, I studied the song: it rolled through three notes forming nearly a major third; the highest note was out of tune and a little low, though not sufficiently to be able to mark it as a [semitone] flat. With respect to the length or duration, these notes were quite equal between them. Only the tonic was prolonged, from time to time, and formed a hold [*tenue*], whose length varied. Consequently, I notated the song as follows:

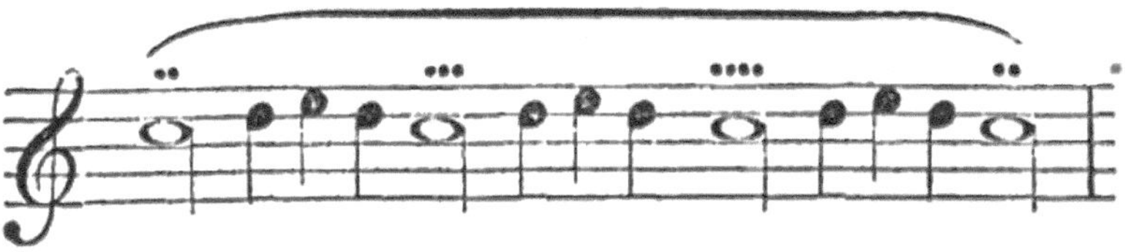

The sound was weak, as if coming from far away, a little shrill [*aigre*] and much like the sound of a guimbarde ["Jew's harp"], with the difference that that rustic instrument can only produce staccato notes [*notes pointées*] and that here all the notes were slurred.[54]

Laennec's stethoscope was the facilitator for hearing the multiple melodies of the heart, and his *Traité de l'auscultation médiate* a guide that attempted to create a system for detecting the different tones and rhythms to help diagnose ailments. Laennec's descriptions, however, were allegorical or functioned by analogy, and the use of an imaginative system led to scientific inaccuracies. Laennec's systematization of sounds was far from accepted as a reliable medical system: auscultation required proficiency in the detection of sounds, grasped only after numerous trials. As such, it was an intuitive and imaginative skill rather than a scientifically proven medical method.[55]

Journals, newspapers, and even the translator of Laennec's first edition in English, Sir John Forbes, questioned the effectiveness of Laennec's technique and found the idea of hearing the body using a simple tube to be ludicrous.[56] Yet there were others more amenable who could glimpse its relevance, like the French writer and historian Françoise-René de Chateaubriand (1768–1848). In his *Mélanges littéraires* of 1819, he foresaw the importance of the instrument in different fields:

> By a tube applied to the outer parts of the body, our learned Breton compatriot, Dr. Laennec, has succeeded in recognizing, by the nature of the sound of respiration, the nature of the affections of the heart and the chest. This beautiful and great discovery will be epoch-making in the history of art. If one could invent a machine to hear what is happening in the consciousness of men, it would be very useful in the times we live.[57]

Chateaubriand's predictions of the stethoscope's impact on the arts soon eventuated. The instrument inspired poets and artists, and its early influence can be found in literature. Take, for example, an anonymous poem published in the *Lancet* medical journal in 1829, which expresses both the impact of the stethoscope on the artistic imagination and the fears the instrument aroused:

> Quoth Rod'rick, "I'll a place contrive
> So dark and safe, no man alive
> Shall to our private meetings grope":
> "Egad," cries Johnny, "that won't do,
> If there's no crack to listen through, They'll take 'reports' by Stethoscope!"[58]

Within Victorian society, Laennec's stethoscope was conceived by many patients not only as a medical instrument by which the cacophonous inner soundscape of the body could be accessed but also as an intimidating device that was capable of revealing the inner secrets of the heart. Patients were anxious about a "curious object" that could hear their thoughts, feelings, and personal affairs expressed through their inner rhythms that they could not control.[59] As Kristie Blair has argued in her study, *Victorian Poetry and the Culture of the Heart*, heart diseases had a prominent role in the medical culture and common life of the time. Widely studied by physicians and discussed in media articles and at public gatherings, maladies of the heart were the subject of general interest. They were attributed to feelings derived from one's circumstances or relationships, and so passionate individuals, such as artists and writers, were considered more prone to heart diseases.[60] This conviction was reflected in paintings and literature. This explains why, with the heart-centered arts production in the nineteenth century, "in a kind of stethoscopic reading . . . the pulse speaks at amplified volume of the sensations it experiences."[61] Thus, instead of being an invisible and silent sign, the pulse in the arts will be shown as something loud and throbbing, and subject to strong emotions such as fear, love, anxiety, shame, and sadness.

The poet Alfred Tennyson, who was very aware of developments in medicine and whose writing style was clearly influenced by the pulse and its rhythms, had a fascination with the heart and the pulse as poetic devices. Heartsickness and the heartbeat were constant themes in his work.[62] Edgar Allan Poe referred to the ubiquitous and uncontrollable throbbing of the heart as a threatening pulse that produces paranoid delusions, as it can be projected externally as a persecuting sound.[63] In "The Tell-Tale Heart" (1843), Poe expresses the desperation of the protagonist who cannot stop hearing the beating heart of the man he murdered: his victim's heartbeats are strong and audible enough to pursue his killer, informing of his crime through their mechanical "loud" and "hideous" sounds.[64] Charles Weste Cope's compelling painting *Palpitations* (1844) depicts a woman anguished over an intercepted love letter. Her anxious expression, in conjunction with the painting's title, reflects a heart under stress, beating strongly when confronted by a compromising situation.

Artists and writers alike had a conception of heartbeats as an indiscreet sign that could reveal innumerable confidential matters.

Despite the fear of invasion into the personal realm and possible inaccuracies in the method, the stethoscope gained recognition and presence in the medical field. It created the innovative possibility of using one object to study another object: the body. By the late nineteenth century, practicing auscultation became integral to everyday medicine, leading to further developments of the instrument. In 1851, the Irish physician Arthur Leared refined the stethoscope from a monaural to a binaural device made of gutta-percha. A year later, the American doctor George Cammann created the first stethoscope for commercial production. It is similar in form to the version that we know today, with two silver tubes and ivory earpieces ending into a single bell-shaped chest piece. Further advances came in the nineteenth and early twentieth century with the invention of electrical stethoscopes and the sound-recording phonocardiograph, which amplified heart sounds more loudly and clearly. By providing recordings and better sound to both ears, these new technologies helped physicians to focus on the internal sounds of the patient, avoid the distractions of external noise, and distinguish normal noises from abnormal ones.[65] They made hearable the language of the heart.

The invention of the stethoscope, which went on to become a symbol of the medical profession, fulfilled the long-held desire to hear heartbeats clearly. In the early 1960s, the development of the Littmann stethoscope, which is still used by physicians today, led to a lighter and more acoustically powerful instrument. When it became more accessible at the end of the decade, more imaginative uses of the stethoscope and the cardiac sounds that connected the inner and outer realm would occur when artists such as Jean Dupuy, Linda Montano, Éliane Radigue, and Christian Boltanski—all practitioners to be covered in future chapters—used the device for collecting and amplifying the movements of the heart. With the aid of microphones or loudspeakers in performances, installations, and experimental sound compositions, these artists mingled art, life, and technology, and listened to the heartbeat in multiple thoughtful and sensitive ways. So, the instrument that gave access to the language of the

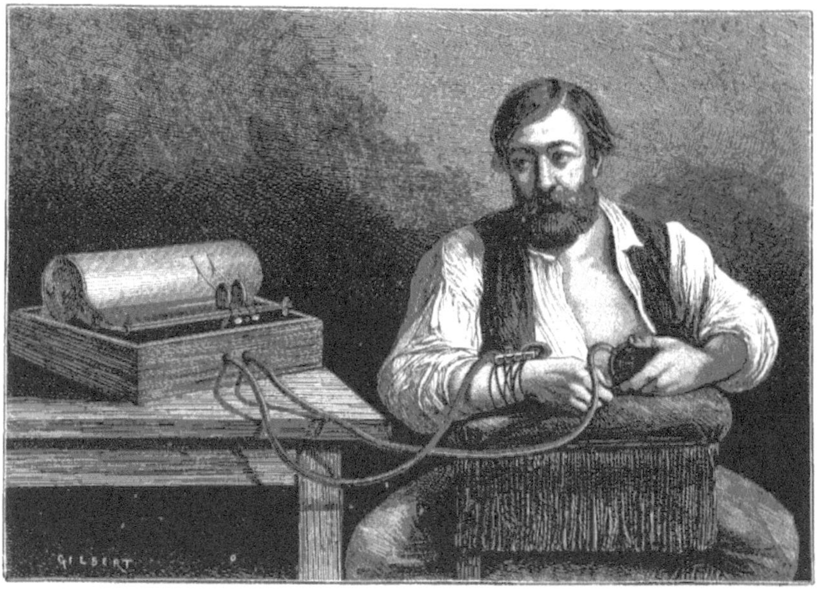

heart became a material that now has a long-standing place in the history of arts.

GRAPHING HEARTBEATS: ÉTIENNE-JULES MAREY'S MACHINES

A man, relaxed and sitting with his arms resting on a cushion on his legs, has an instrument fixed to his wrist that is connected to a rubber hose that leads to a machine on a table. This metallic rectangular apparatus is a polygraph drawing a zigzag line on paper that, while generating a mechanical ticking sound, records the subject's vital signs—namely, his breath and pulse, and, therefore, his life. The illustration in figure 1.12, painted

1.12

Portrait of Étienne-Jules Marey by Gilbert, *La Nature*, 1879.

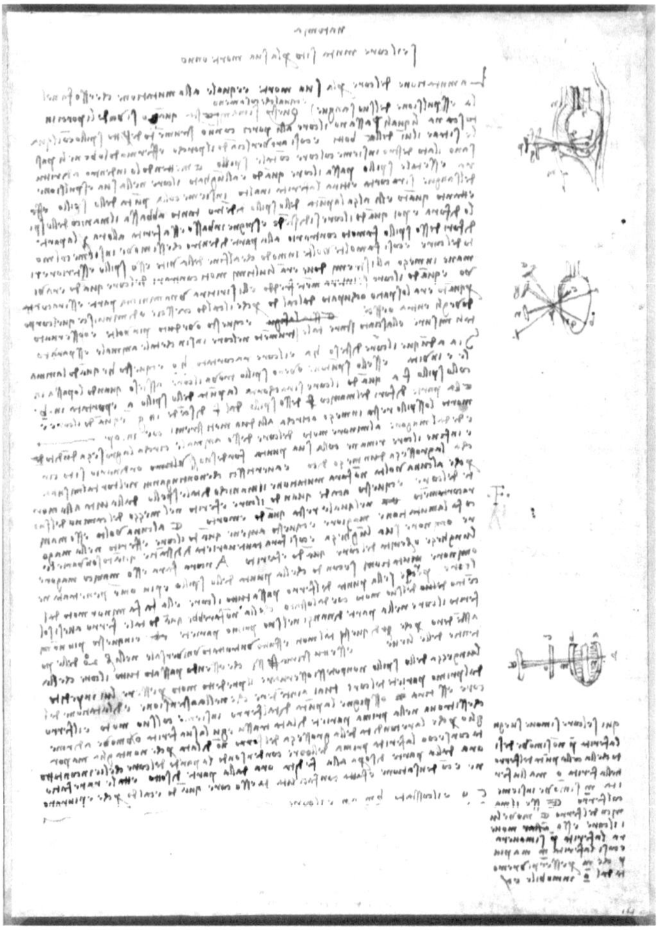

1.13

Two of Leonardo da Vinci's drawings of the motions of a pig's heart, ca. 1511–1512. © Royal Collection Enterprises Limited 2024 | Royal Collection Trust.

by Gilbert in the 1870s and published in the French scientific magazine *La Nature*, is a twin portrait of the French physician Étienne-Jules Marey, representing both the subject's body and the body as a linear, mechanical self-inscription. In the image, both machine and person are connected by the heart and its circulatory system, but also through cords and canals, which results in a physical and diagrammatic scientific representation of the body in a single space.

The visualization of the pulse, that bodily vibration hiding under the skin, rarely perceptible but constantly operating, was passionately pursued by physicians for a long time. To be able to see what could not be draw out required some imaginative and innovative methods. Galen, in the second century, for example, realized numerous insightful visual experiments, such as carefully placing a hollow reed within the lumen of an artery so as to see and analyze the transmission of the pulse.[66] Leonardo da Vinci studied the cardiac movements of the pig's heart and created one of the first known anatomical drawings to convey this knowledge (figure 1.13).[67] William Harvey repeated Galen's experiment using a ligature to demonstrate that the heart valve's function was to produce the unidirectional flow of blood (figure 1.14).[68] This investigation led him to discover the circulatory system. The Polish physician Josephus Struthius developed a number of enchanting visual exercises, such as placing a leaf on the artery of a patient to watch it vibrate in time with the pulse wave.[69] He also published the earliest graphic depiction of the pulse in his book *Ars Sphygmica* (1540), in which he drew two lines, one curved and another jagged, representing two different kinds of pulses.[70]

However, the direct visualization of the pulse became possible in the nineteenth century by the use of a series of clockwork apparatuses and a stylus for marking it as a continuous line. These apparatuses, many of which involved testing in animals, were part of a series of inscription technologies—such as phonography and photography—that recorded forms of energy (e.g., sound, heartbeat, or light) onto paper or glass, creating a memory of the ephemeral.

One of the first pulse wave recording apparatuses was built in 1846, when the German physiologist Carl Ludwig (1816–1895), moved by his interest in studying the relation of respiration and blood pressure rates, created the first instrument that graphically traced the variation of blood

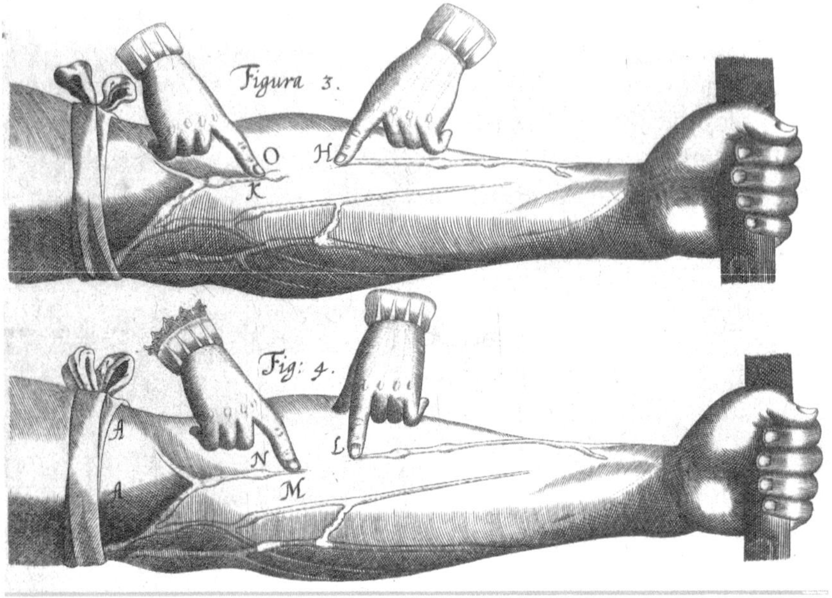

pressure: the kymograph. This intrusive "wave writer" apparatus operated through a metal cannula, inserted into the artery of an animal, as a lever that transferred the contractions of the heart to a stylus that recorded changes in arterial pressure. While this visualization of the continuous wave of the blood flow had a great impact, the instrument proved too invasive for clinical application.

It was a decade later, in 1855, when the physiology professor at the University of Tübingen, Karl von Vierordt (1818–1884), after many experiments, created the first noninvasive device that permitted the arterial pulse to be monitored and drawn, as well as studied as a wave. The sphygmograph consisted of a system of weights and levers that estimated the blood pressure rate. Vierordt used his sphygmograph with numerous patients, families,

1.14

William Harvey, *Exercitatio Anatomica de Motu Cordis et Sanguinis in Animalibus*, 1628. Wellcome Collection.

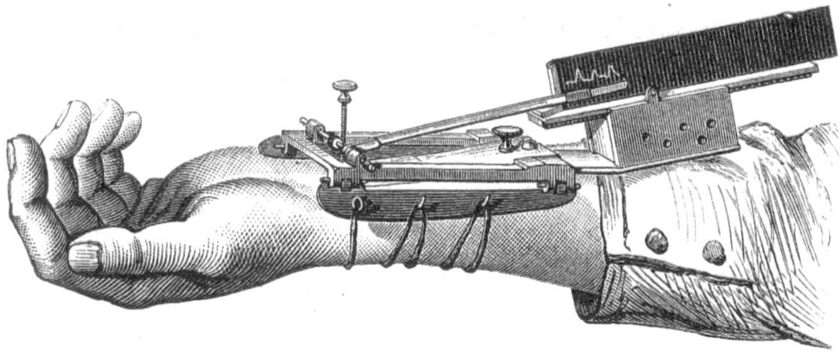

and students, and these experiments are recorded in many delightful steel plates. Some are beautifully poetic, such as the first pulse recording that was that of his wife, while others record the pulse in different states of life and death. Vierordt's instrument, however, lacked accurate information on the rhythm, frequency, and amplitude of the pulse and was considered limited and inexact.

A few years later, in 1860, the French physiologist and photographer Étienne-Jules Marey (1830–1904) improved Vierordt's instrument by making it portable and able to generate an accurate graphical record of the pulse wave. Marey's sensitive mechanical device, which allowed a view of the heart-writing, was simple but very innovative for that time: it consisted of a mechanism fixed to the wrist with a double-ended lever. One end connected to a stylus and the other to an ivory sensor that was placed against the radial artery of the wrist (figure 1.15).[71] A clockwork mechanism moved the stylus over blackened paper, into which the wavy rhythm of

1.15

Étienne-Jules Marey's sphygmograph in use, 1885. Wellcome Collection.

the pulsation was etched, giving form to the throb that keeps the body in motion.

The feasibility of Marey's sphygmograph made it a rapidly accepted and accessible tool in clinical medicine. Nevertheless, the most important feature of the machine for Marey was its reliability based on a graphic method. This system revolutionized not only the epistemology of heart observation but also how the body was perceived. The emotional, mystical, and religious principles that characterized medicine before the invention of instruments can be clearly seen in plates of work by medieval physicians like Robert Fludd. In Fludd's book *Pulsus*, for instance, a hand that is measuring the pulse emerges from the clouds, representing the influence of the universe and the cosmos on human health. Before the sphygmograph, physicians described and studied the movements and sensations of the pulse using allegories or analogies—such as strong, weak, anting, worming, or gazelling.[72] These systems, however, based on mystic and subjective perceptions, proved problematic. With the sphygmograph, this abundance of imaginative and metaphorical descriptions, all subject to variable appreciations, was replaced by a curve. Even though it relied in fact on a combination of formal and intuitive readings, it was accepted as an "objective" technique.

Regarded as impartial, the graph displaced the "listening" model that was predominant at the beginning of the nineteenth century with an illustrative model of diagnosis. In this opto-centric system, sound, nature analogies and imagination were dismissed insufficient and inferior to visual rationality. In his *La méthode graphique dans les sciences expérimentales et principalement en physiologie et en médicine* (1878), Marey describes graphic lines as a form distinct from language, which was "born before science and not having been made for it."[73] Created by the swinging movements of a stylus without human intervention, the graph was the "language of phenomena themselves," the solution to "the insinuations of eloquence and the flowers of language,"[74] and certainly superior to the misleading senses:

Not only are these apparatuses destined sometimes to replace the observer, and in their circumstances fulfil their role with an indisputable superiority, but they also have their own domain, and nothing can replace them. When the eye ceases to see, the ear to hear, and the sense of touch to feel, or when our senses deceive us, these apparatuses are like new senses of astonishing precision.[75]

Clearly, Marey considered both the senses and palpation to be fallible, and that the instruments used until then could not compare with the precise data provided by the sphygmograph. The stethoscope, for instance, was an instrument that could help extend the senses, but it did not provide objective data about the heart. With it, doctors could certainly detect change and give a subjective description of the pulse rhythm, but the sphygmograph offered so much more. Marey was capable of providing a visual record of the variations of blood pressure and the pulse (figure 1.16), and the graphical recordings were not only able to make the previously indefinable and imperceptible arterial pulse visible but also to make it understandable to the eye and the brain.[76] As Martha Braun has noted, Marey's instrument helped to confirm Harvey's theories that the pulse was produced not by the expansion of the heart but by its contraction, which was still an important unresolved topic at that time.[77]

With the intention of extending his knowledge of vascular circulation, Marey expanded his visual experiments with the sphygmograph through a series of collaborations with the recognized veterinarian and physiologist Auguste Chauveau (1827–1917). Chauveau encouraged Marey to expand his interest to include cardiac motricity, and together, in addition to publishing papers in specialized journals, between 1861 and 1863, they created an innovative technique: the cardiograph. A complex and invasive device at that time used only for horses, the cardiograph consisted of three sphygmographs that connected a set of styli to a foal's heart chambers via a number of catheters. The styli traced lines reflecting the movements of the atrium, the ventricle, and the heartbeat on a sheet of paper wrapped around a rotating cylinder. "By looking at the line of the external beating of the heart," Marey declared, "it was possible to know everything that was happening within this organ."[78] While the stethoscope revealed the secrets of the heart to a doctor, the heart's writing spoke directly to anyone who wanted to read its inner mysteries. This universal and sincere language came with its own risks, which can be clearly found in another communicative heart instrument made by Marey: the polygraph.

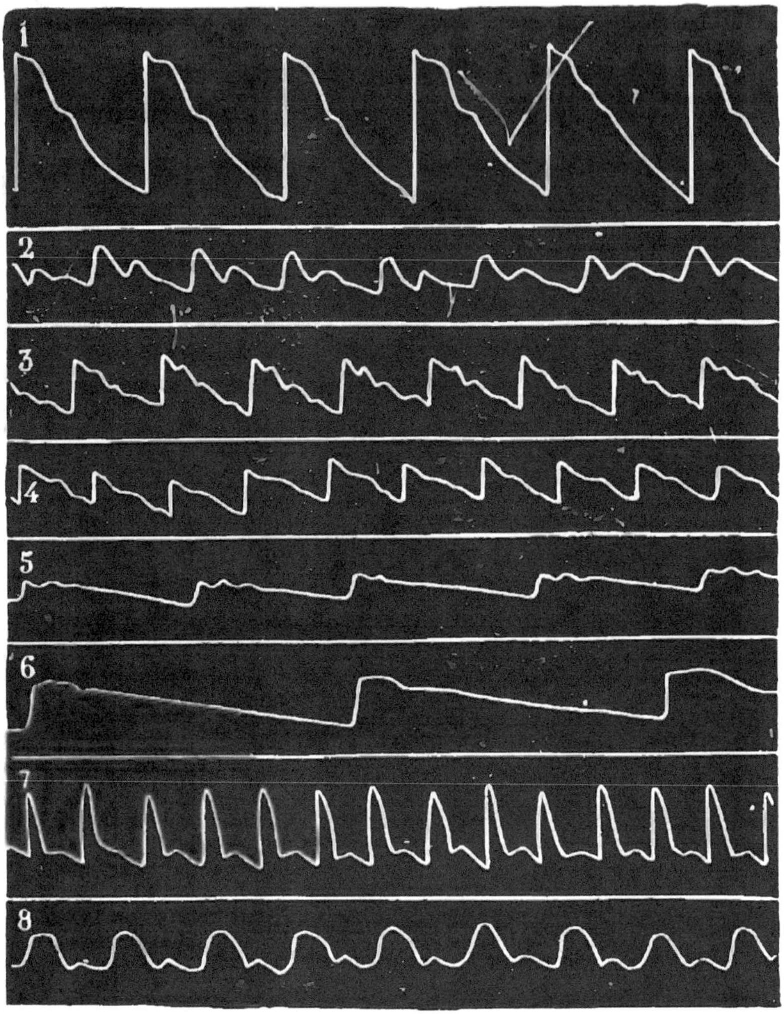

1.16

Étienne-Jules Marey's enlarged tracings of the pulse in different diseases, 1895, from Étienne-Jules Marey, *Movement*, trans. Eric Pritchard (New York: D. Appleton and Co., 1895).

Like many physiologists of the time, Marey was obsessed with creating instruments to visualize and record physiological functions. Among the instruments he created, the polygraph was outstanding because it was not only able to scan the heart and its secret variables but also those of the entire body—pulse rate, heart and artery pulsation, respiration, and muscular contraction.[79] With these instruments, Marey created a new language to decipher the body by registering all its movements. This produced an information system of converging energies that has never been seen before. Further, he established the basis for an instrument that a century later will speak even against the heart itself in judiciary processes.

The sphygmograph, the cardiograph, and the polygraph were some of the many apparatuses that Marey invented during his lifetime, all of which attempted to reveal and record barely perceptible cardiophysiological phenomena. The creation of these devices occurred at a time when energies—specifically in theories on conservation, thermodynamics, and bioelectromagnetics—were at the center of scientific discussions. Marey's interest, however, was focused on a primary form of energy: movement. Marey's fascination with mobility and visualizing it in space and time started with blood circulation but later expanded to human and animal locomotion. The flight of birds, the muscular contractions of a frog, and the motions of an animal's heart all included slow as well as fast transient movements that Marey wanted to capture and analyze. For this purpose, in 1882, Marey developed an electric photographic gun using 35 mm film, which could produce twelve images per second on a turning plate. Marey called this technique, which anticipated cinematography, chronophotography (figure 1.17). As its name indicates, this was a graphic inscription method producing a series of images of bodies in motion but fixed in time and space. To study the perception of time in movement and to prove that in short intervals, many things remain unperceived, Marey used dark backgrounds. By strategically attaching strips of white paper to the bodies of his subjects, he could capture the minutiae of movement in a set of marks or "skeletal trajectories." Photography was a self-inscriptive technology—a mode to record life—and Marey used it as a graphical method to capture movement, which he considered the most important characteristic of life.

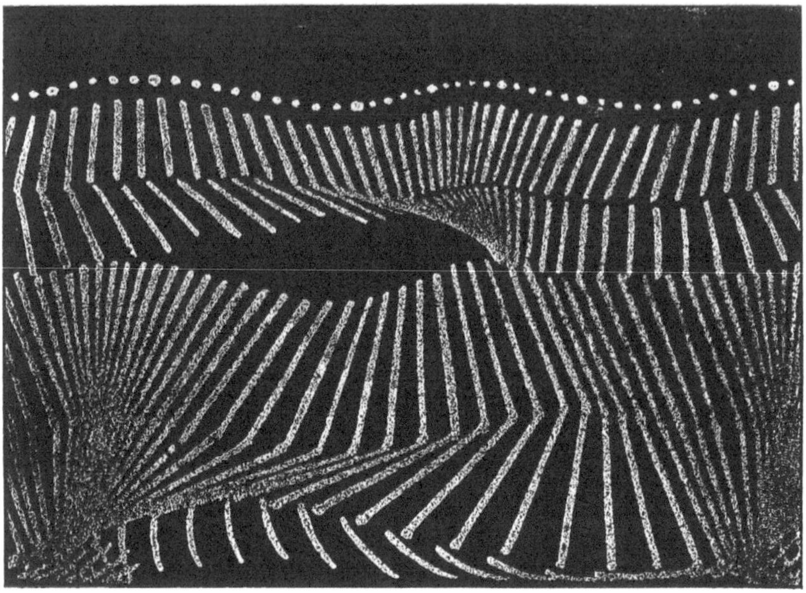

With the intention to "elucidate what the unaided eye cannot discover for itself," Marey created a series of seven successive photographs using the chronophotographic method. His goal was to see the movements of the heart of a tortoise under conditions of artificial circulation (figure 1.18).[80] This important experiment, described in his book *Movement* (1894), was developed from a simplified version of a previous model. A glass funnel was introduced into the tortoise's left auricle, and a tubing system—made with glass cannulas and rubber tubes—acted as an artery connecting the aorta and the glass funnel through which blood was poured.[81] Marey tainted the organ with white paint to enhance the

1.17

Étienne-Jules Marey's chronophotograph of a runner, 1880, from Étienne-Jules Marey, *Movement*, trans. Eric Pritchard (New York: D. Appleton and Co., 1895).

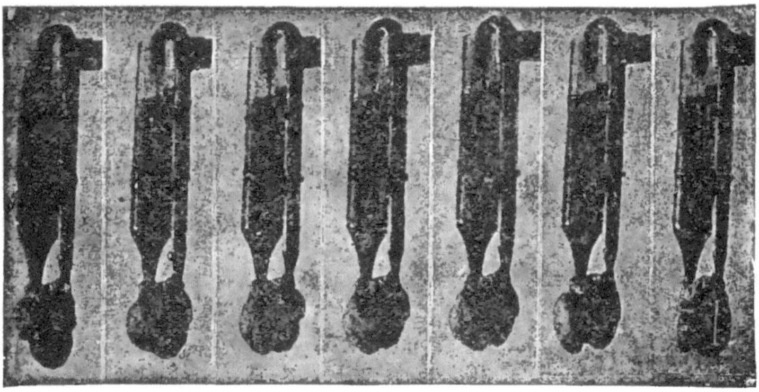

movements of the heart so he could have a better appreciation of beating and show its constant contractive and expansive movements. These mechanical structures were a recurrent methodology Marey used to study the cardiovascular system. Among the many models he created were an artificial heart made with rubber and valves, and a complex model of the circulatory system that used cords, pulleys, and a flywheel, among other materials. These instruments, intended for simulating cardiac activity, aimed to "reproduce the form of the heart beat and of its variations."[82]

For Marey, observing the heart's activity was indispensable to understanding its function, and so he reproduced the body outside the body via apparatuses that became the body itself (figure 1.19). Marey's artificial body systems were more than a simple curiosity or entertainment; they were sophisticated devices created by experimental trial and error that provided the opportunity to see and hear the invisible movements of the living heart.[83] His sphygmographic plates are portraits of life that bring art and physiology together. Marey's chronophotographs of the turtle's

1.18

Étienne-Jules Marey's chronophotographs of a tortoise's heart with artificial circulation, 1895, from Étienne-Jules Marey, *Movement*, trans. Eric Pritchard (New York: D. Appleton and Co., 1895).

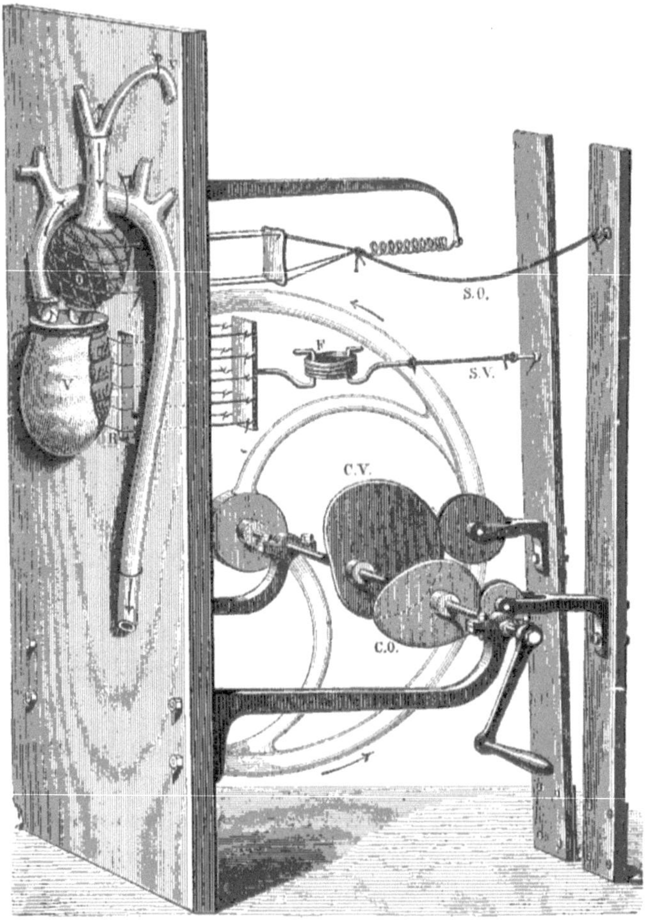

Fig. 3. — Schéma de la circulation du sang.

1.19

Two of Étienne-Jules Marey's models from *La Nature* 127 (1875): 357–358. This page: Machine for imitating circulation. Facing page: Artificial heart model for reproducing valve sounds.

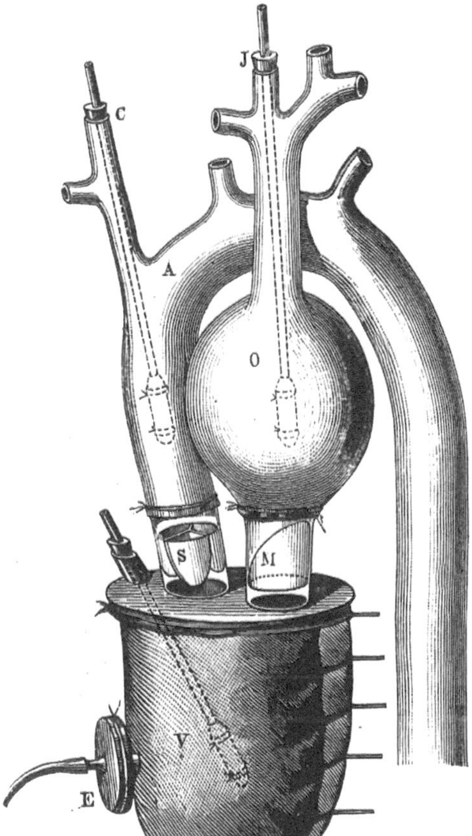
C
J
A
O
S
M
V
E

heart can be considered the first moving images of a pounding heart, while his models of artificial hearts can be thought of as among the first machinery or "installations" created to make visible to the eye the rhythmic movements of heartbeats.

The visualization of body movements guided Marey's studies, but his interest in mobility did not stop there. He also researched the sounds produced by heart movements. This interest, which has received less scholarly attention, is not surprising since sound, as a wave, is also a form of movement and an indicator of life. Like his predecessors, Marey linked the rhythms of bodily sounds to musical tones, but he used a simplified system based on simple descriptions and charts. As a rhythm, Marey recognized in the heart a pattern of strong and weak measurable movements that were accompanied by two noises: a dull and grave "premier bruit" that was accompanied by "an energetic shock of the whole heart,"[84] followed by a clear and audible second noise. Between these two noises were two silent intervals, one small and one long, each of variable duration. Considering that the music notation techniques used by previous researchers to represent the rhythm of the noises of the heart were not uniform, Marey moved from music notation to what he called an "analogue expression"—namely, a chart that used lines and the names of the sounds.

With this graphic method, Marey also found a way to depict cardiac sound activity in a more simple and understandable system. His book *La circulation du sang à l'état physiologique et dans les maladies* (1881) contained a diagram graphing the relationship between the sounds and the movements of a horse's heart. A jagged continuous line registered heart sounds while wavy lines recorded the force of the blood pressure. In this way, Marey showed the intensity of the heart's movements displayed in a horizontal successive coordinate graph. Exactly as in music, this graph portrayed the rhythms as temporal and spatial displacements. Furthermore, for analyzing heart noises, Marey created a machine for artificially reproducing the sounds of the valve. The model consisted of a rubber bulb connected to tubes that pushed air into the "heart," generating a systole-like motion accompanied by a sound. Through aesthetic prostheses, graphics, and machines, Marey reproduced the phenomena of the heart outside the body, opening a novel avenue to see the mechanisms of the invisible and to understand their language. His work would go on to have an impact on the arts.

Marey's models and self-recording instruments were, as Robert Brain has stated, "practical means of deterritorializing the body." They opened up new spaces—that is, as an inscription on paper or a trace of the body made with a mechanical machine.[85] The graph, which was a system of analysis but also of distance and control, had an impact on modernist aesthetics that can be traced to anthropometry. This study of the measurement of the human body as a recognition system, based on the use of indexed physical features, was used to identify recidivists. To recognize criminals, French police officer and biometric researcher Alphonse Bertillon (1853–1914), who was a contemporary of Marey at a time when graphs were a topic of discussion in scientific circles, used anthropometry cards with full-face photos and mugshots of criminals along with detailed measurements (figure 1.20) based on eleven body parts (heads, arms, ears, feet, etc.), which were attached to photographs organized in a mathematical file based on a graph or binomial curve.[86] Like Marey, Bertillon devised the photographs with body measurements and a system of graphs as a way to "educate the eyes" to see something that was not immediately visible without access to the collected data.

In the arts, authors like Robert Brain, Martha Braun, and François Dagonet have shown that Marey's graphs and chronophotography, although created with purely scientific intentions, had an enormous and very different later impact on the development of cubist, surrealist, futurist, and kinetic art aesthetics. Although graphs had a considerable impact on scientific thinking and were considered an objective and nonsymbolic language, artists would later be in charge of proving the opposite. Successive generations of modern artists found in Marey an inspiring figure. Seeking to separate themselves from traditional aesthetics, artists looked to science for something that was more in tune with the technologies and machineries of the industrialized society they were living in. Marcel Duchamp's *Nude Descending a Staircase No. 2* (1912; figure 1.21) and Giacomo Balla's *Study of a Girl Running on a Balcony* (1912; figure 1.22), both portraitures tracing the displacement of bodies in space, demonstrate

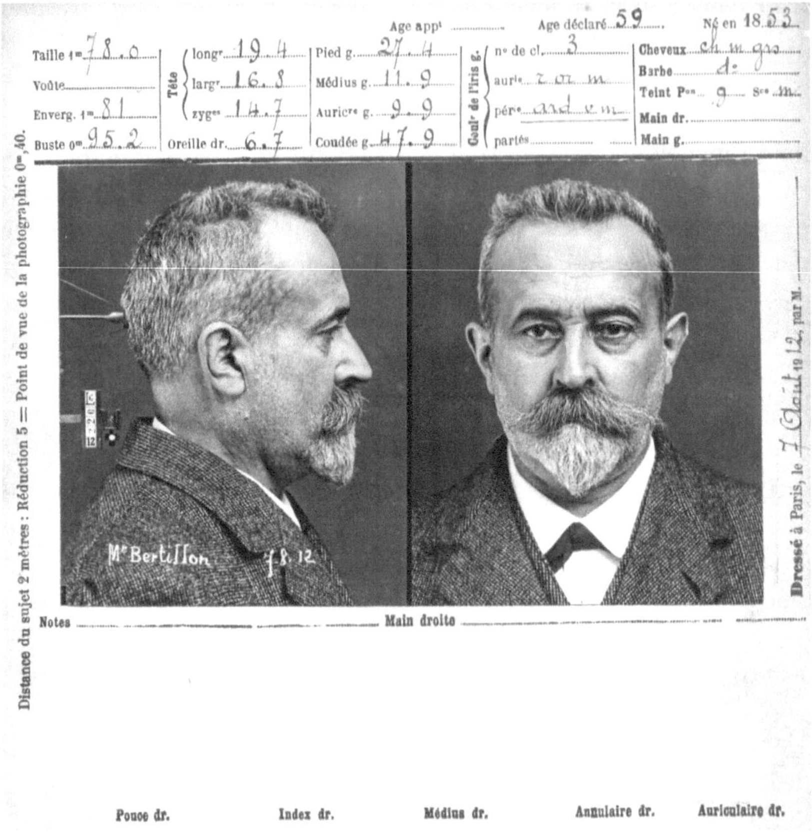

1.20

This page: Anthropometric data sheet of Alphonse Bertillon, 1912, Paris, Archives of the Police Prefecture. Facing page: Binomial Curve of Distribution (Depicting Height in French Males), 1896, from Alphonse Bertillon, *La photographie judiciaire* (Paris: Gauthier-Villars et fils, 1890).

hauteur est proportionnelle au nombre de sujets de cette taille que l'on observerait dans une population de 1000 adultes, du sexe masculin.

L'expérience a prouvé que tous les pays du monde présentent une concordance presque rigoureuse entre la courbe théorique donnée par la formule mathématique et celle qui résulte de l'observation.

Le chiffre de la taille moyenne ou centrale

Fig. 6.

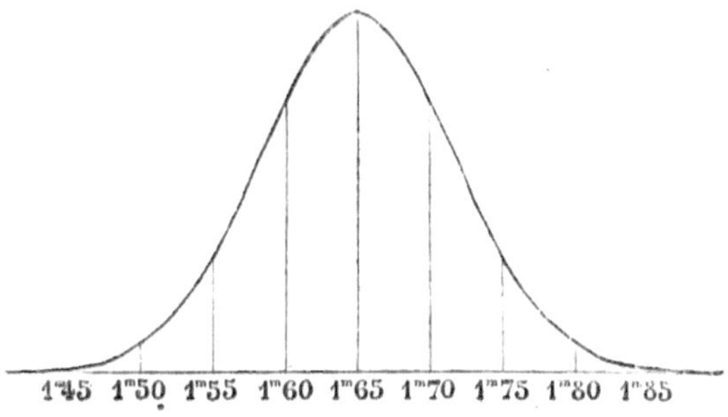

Courbe de probabilité de la taille masculine en France.

change suivant les races; mais, une fois sa valeur connue, la fréquence de toutes les autres tailles s'en déduit et la courbe de probabilité (*fig.* 6) prend universellement la même forme.

Cette curieuse loi de distribution est observée par la nature, non seulement pour tout ce qui con-

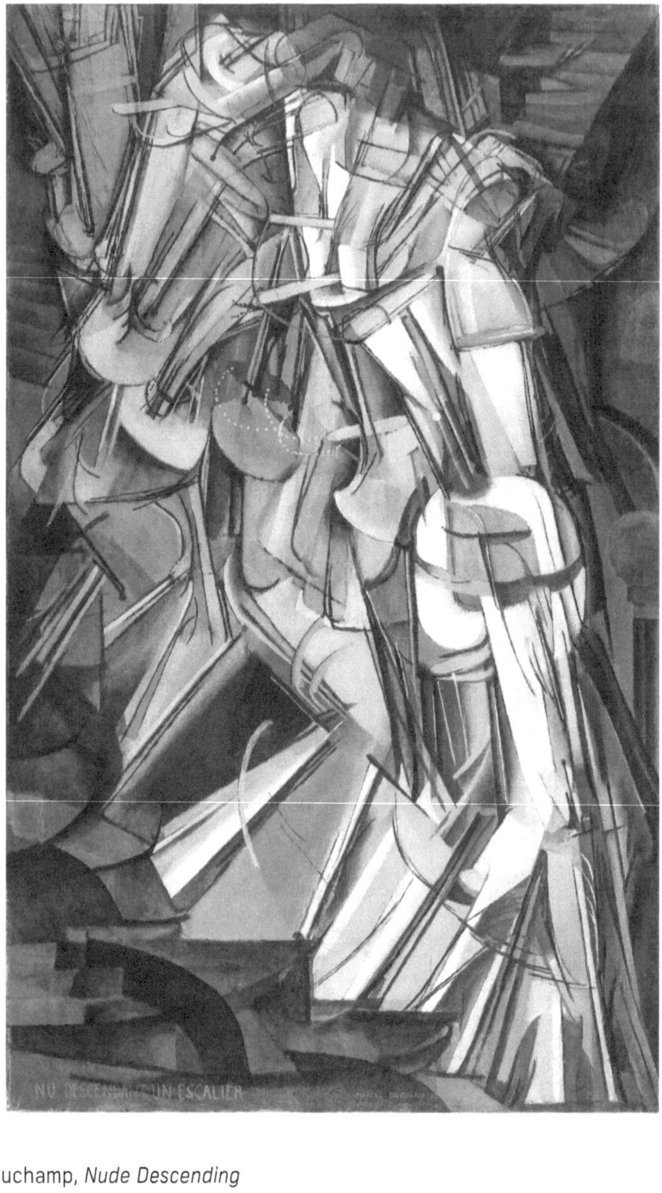

1.21

Marcel Duchamp, *Nude Descending a Staircase No. 2*, 1912. Courtesy of the Philadelphia Museum of Art. © Association Marcel Duchamp / ADAGP, Paris / Artists Rights Society (ARS), New York, 2024.

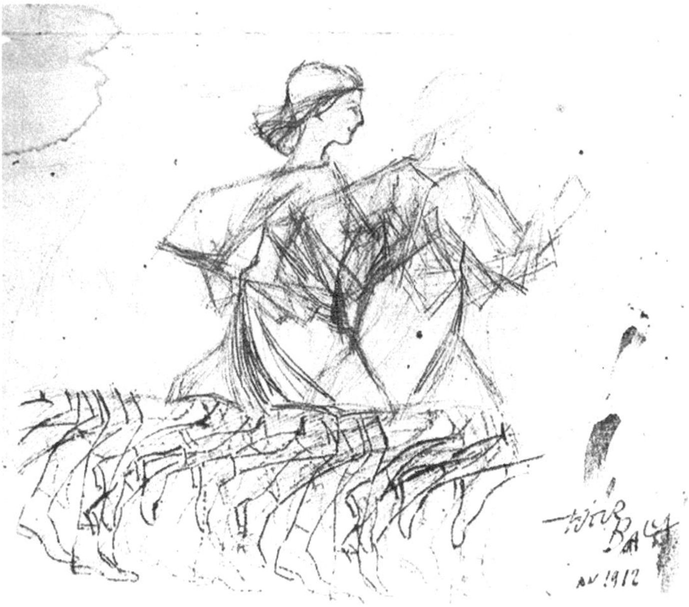

how Marey's chronophotography had an impact on the arts of the early twentieth century.[87] Graphic instruments also influenced aesthetic theories, such as Charles Henry's *esthétique scientifique*. It attempted to adapt experimental physiology to the arts and to make curved lines the base for a new aesthetic language that yearned for a rhythmic and machinist conception of art.[88] The surrealist Max Ernst subverted images from *La Nature*, the popular nineteenth-century science magazine, including some images of Marey's scientific experiments, with the intention of creating contradictory collages that "make visualizations of the invisible within himself."[89] Inspired by science, Ernst also started to include graphs and

1.22

Giacomo Balla, *Study of Girl Running on a Balcony*, 1912. © 2024 Artists Rights Society (ARS), New York / SIAE, Rome.

diagrams in some of his drawings in *Misfortunes of the Immortales*, a publication of collages and poems produced in collaboration with Paul Éluard.

Marey's images and models still have an impact on the aesthetics of contemporary artworks today. While the sphygmograph sparked the curiosity of artists to create works that combine the language of science and physiology, the instrument that provided them with heart records a century later was not Marey's sphygmograph. Another technology had overshadowed it, as it was able to record the electromagnetic energy of the heart: the electrocardiograph.[90]

ELECTRICAL HEARTBEATS: WILLEM EINTHOVEN'S ELECTROCARDIOGRAPH

A group of elegantly dressed men surround a dog on a table. Two of its legs are immersed in glass jars containing a saline solution and then connected by wires to a light bulb and then to a string galvanometer, which is a thin wire thread that conducts an electric current that is stretched between two points in a magnetic field (figure 1.23).[91] In the engraving illustrating the event, published in *The Illustrated London News* in 1909, the men are attentively observing the canine's heart via electric currents on a screen. On the far left is the British physiologist, A. D. Waller, who is demonstrating Willem Einthoven's galvanometer on his bulldog, "Jimmy," at the Royal Society in London. Einthoven's electrocardiograph was the first noninvasive, light-dependent, and cinema-based heart tracing device. A powerful arc lamp illuminated a quartz filament, whose varying movements were projected as shadows onto a screen, showing the electrical variations of the heart. These continuous oscillations were a visual delight that fascinated audiences, and many of them allowed their personal heartbeats to be projected in an exterior public space.[92]

Events and demonstrations such as these—where medicine, art, science, and physiology intersected—used various elements of cinema experimentally, such as "storage and reanimation, projection, lensing, modulation of light, and artifice in a darkened space."[93] The result was spectacles that attracted a broad range of people—including women who were present but excluded from accounts—all of whom found the

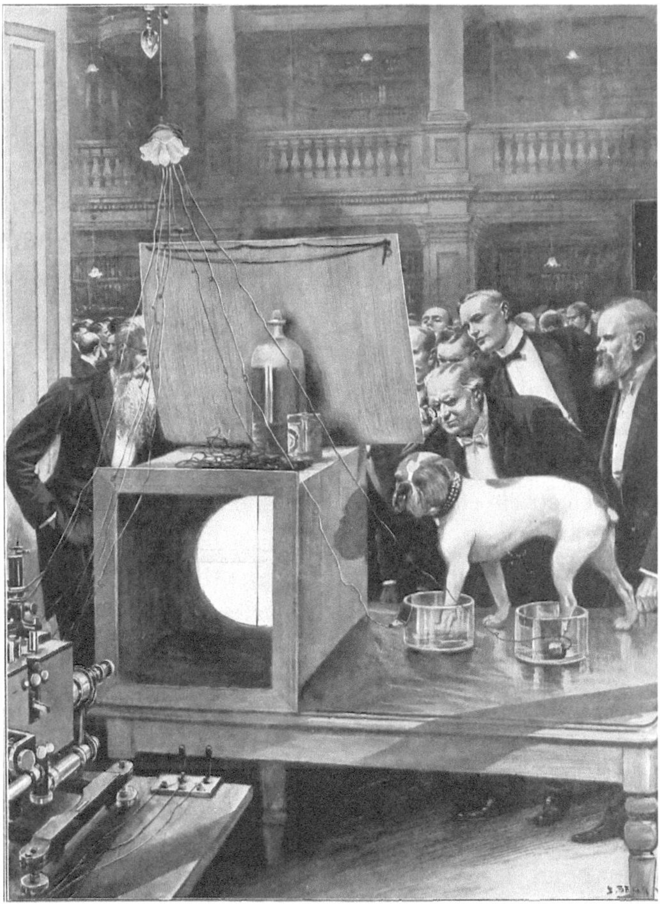

1.23

The physiologist A. D. Waller at the Royal Society demonstrating the electrocardiograph on his bulldog. *The Illustrated London News*, May 22, 1909.

movements and sounds of the heart alluring when made visible through its electrical impulses in electrocardiograms. Unlike previous experiments I mentioned that involved the bodies of animals, Jimmy's demonstration was also important because, although an unpainful procedure, it sparked a debate in the British Parliament about whether the Cruelty to Animals Act, which had passed a few decades before in 1876, had been violated.[94] The intent audience in figure 1.23 testifies to how the electrocardiograph and its ability to register cardiac activity has inspired creativity, entertainment, and social discussions since its earliest beginnings.

This enthusiasm for turning visible corporeal electrical phenomena into public performances can be traced to the eighteenth century. The key elements were usually devices or mechanical instruments connected to bodies in darkened spaces. Most of them were based on connections between the electrical conditions of natural phenomena, electrical machines, and the body itself. Scientists like Benjamin Franklin found similarities between lightning in the sky and the sparks created by their electrical apparatuses and prompted questions about whether that electrical activity could also be found in the body.[95] Once such activity was found, Franklin and other famous physicists like Jean-Antoine Nollet toured Europe, giving demonstrations of the attraction and repulsion of charged bodies and of electrical sparks by means of an apparatus.[96] Nollet presented the "Leyden experiment," in which a display of very powerful shocks and sparks provided by a Leyden jar filled with water hung from the collector of an electrical mechanism by a metal clasp.[97] By touching the inner and outside surfaces of the jar with both hands, it provoked an electrical discharge on the body that intrigued audiences. Like many physicians, Nollet gave an aesthetic importance to the creation of his instruments for attracting audiences:

> A pretentious apparatus may be used with great effect to show phenomena that are already known. I have great regard for the elegance of instruments, which now adorn our schools and the cabinets of amateurs. The facts discovered there owe nothing in their certainty or in their utility to instrumental decoration, but when these facts are shown more gracefully it may be hoped that they will evoke greater interest.[98]

Such spectacles were the product of systematic research during the eighteenth century, culminating in the most significant achievements in the nineteenth century. Displays of dramatic electric sparks and discharges were particularly celebrated in Europe and America in the nineteenth century. The Serbian-American engineer, physicist, and showman Nikola Tesla, for example, created "flame" shows produced by the ionization of gases. In this period, scientists and audiences were captivated by seeing, harnessing, and developing the power of electrical activity and its sparks.

Electricity was regarded not only as a "spectacle" but also as the engine of "modern civilization." From the discovery of electric current by Michael Faraday in 1831, the invention of the light bulb by Heinrich Göbel in 1854, the further developments of incandescent light by Thomas Edison in 1879, and the introduction and diffusion of AC electric motors by Tesla in 1888, the march of electricity was rapid. Electric light became a synonym for "ideas" and "reason" via its possibilities for technological and political development. "Electricity is life" became a popular catchphrase in the period that the electrical heartbeat became visible and new lighting capabilities allowed virtually everyone to transcend the natural cycles of light and dark.[99]

The visualization of the electromotor activity of the body was first observed in small animals that involuntarily gave their lives for the advancement of science. In the second century, Galen mentioned that the heart of an animal removed from the thorax continued to throb for a considerable amount of time, which was a "definite indication that it does not need the nerves to perform its function."[100] This point was supported by Aloysius Luigi Galvani (1737–1798), professor of anatomy at the University of Bologna, with his experiments with frogs in 1787. He accidentally discovered that the nervous termination of the croaker muscle discharged electrical currents when carefully touched with a metal scalpel and saw involuntary movements. This led Galvani to affirm that living tissues had electrical properties, an idea that was disputed by Alessandro Volta (1745–1827), physics professor at the University of Pavia. He denied the existence of animal electricity and invented the "voltaic pile" to show that contractions of the frog's legs were produced by the contact of metals and moistened substances.[101] Although Volta and Galvani were creating competing theories, they were both correct: electricity can be generated by the contact between metals, and electric currents are produced within living organisms, even though apparatuses at that time were not yet able to register this. Thus, some physicians of the eighteenth century, like Albrecht von Haller, based on his observation of frogs, affirmed that the movements of the muscular organ were spontaneous and independent of nerves, while others, such as the English physiologist Thomas Willis, argued that nerves stimulated the heart to contract.

During the same years that Mary Shelley's Frankenstein (1818) came to life thanks to an electric current, giving rise to the first science fiction novel inspired by Shelley's knowledge of the emergent field of electrophysiology, science myogenic and neurogenic theories of the origin of the heartbeat dominated cardiac research. In 1825, the Italian physicist Leopoldo Nobili developed a sensitive galvanometer capable of measuring small amounts of electricity in living tissue. This technology helped Nobili's student, the physicist and neurophysiologist Carlo Matteucci, to observe in 1838 that electrical activity was also found in the heart muscle of a frog. This observation was later substantiated by the physiologists Albert von Kölliker and Johannes Müller in 1856 when they proved that by using animal tissue as a galvanometer, a frog's heartbeats were always accompanied by an electric current.

Today, we are certain that it is the heart's electrical system that triggers contraction. The heartbeat begins in the sinus node, a group of cells located in the right atrium of the heart, which sparks electrical signals that spread to the atrioventricular node. This in turn sends the message to the ventricles, the major pumping chambers of the heart, to contract, which then pump blood throughout the body.

The first accurate visual observation of the electrical properties of the heart is attributed to Dutch physician Willem Einthoven (1860–1927). Inspired by Waller's first recording of the electrical activity of the heart and his experiments with his dog Jimmy, as well as the electrophysiology studies by the Japanese physiologist Sunao Tawara—who discovered the atrioventricular node—Einthoven created the first accurate electrocardiograph. The machine, created in 1902 after many prototypes, translated and recorded the electrical activity of the heart in each systole and diastole as a continuous curve, making visible the invisible energy of the heart. Contrary to today's lightweight and portable EKGs, Einthoven's device

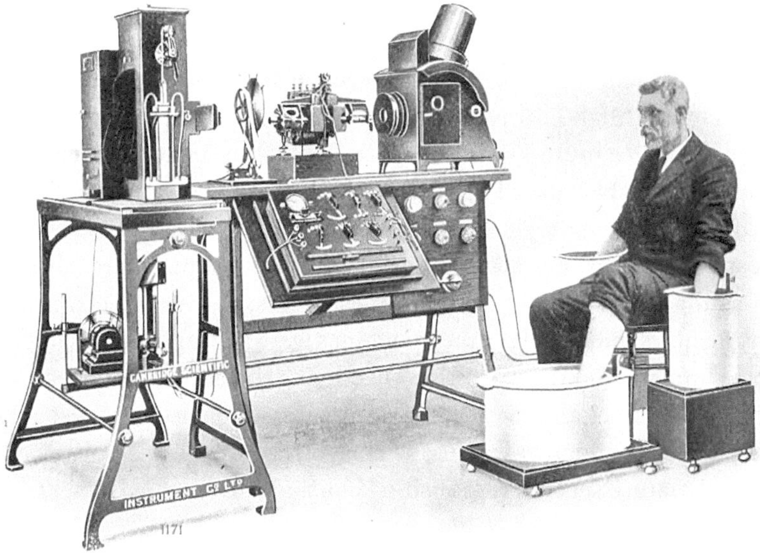

was an immobile and cumbersome machine occupying two rooms and weighing around 270 kilograms because of its sizable electromagnet.[102] The machine was bulky, but it was also operationally complex (figure 1.24). It required at least four people to perform an electrocardiogram, as well as a continuous flow of water because it overheated easily. Its operation started when a person's right hand and left foot were placed in containers filled with a salt solution connected to a mercury capillary electrometer. When a current passed through the conductive wire, the string galvanometer moved, creating a shining light that was recorded onto photographic paper, leaving a trace of the continuous heart waves.

1.24

A table model of an electrocardiograph manufactured by the Cambridge Scientific Company of London in 1911. *The British Medical Journal*, March 25, 1950 (London: The Royal Society of Medicine Press Ltd., 2003).

To be able to see the electromagnetic activity of the heart was an outstanding discovery. In 1924, Einthoven was awarded the Nobel Prize in recognition of his invention of the electrocardiograph. In the award ceremony speech, the chairman of the Nobel Committee, Professor J. E. Johansson, highlighted the importance of the instrument and likened the heartbeat to a brute sound of industrialism: the "piston movement of a steam engine" because of its "cyclic process."[103] This analogy is significant because it mirrors, in the late nineteenth and twentieth centuries, the conception of the heart as a machine, sharing some characteristics, such as the integration of energy, tubes, valves, and rhythms. Furthermore, it shows that, at that time, machines were an important metaphor for how humans saw the world. By objectifying the body as a mechanistic controllable object and by focusing on its functional aspects instead of seeing it as part of natural cycles, science placed the body's holistic connection to nature second, establishing a distance from it.

MECHANICAL HEARTBEATS: FUTURISM, WEIMAR ART, SURREALISM

The development during the nineteenth century of physiological devices—such as the stethoscope, the sphygmograph, and the electrocardiograph—produced detailed descriptions of the functionality of the heart and strengthened the conception of the body as a machine. This idea found its roots in Julien Offray de La Mettrie's *L'Homme Machine* (1747). He extended the materialistic philosophy of Descartes, who had conceived of bodies as functional structures dependent on energy and automatic functions like the constant beating of the heart. The notion of humans as machines and body parts as apparatuses was not new. Centuries before, William Harvey observed the mechanical character of cardiac activity when he compared the heart with a pump and heartbeats with a gun firing. By revealing the mechanisms of the body with its possibility of deconstruction, analysis, and repair, physiological apparatuses fostered the conception of the body as an organic but technological instrument composed of several parts, each with a specific function and task.

Mechanistic interpretations of the human body, and specifically of the heart, proliferated in the early twentieth century when the ubiquity of machines (like the steam engine or the automobile) was having a strong impact on society. Scientists directly related the "cyclic process" of the heartbeat with machines, first with the steam engine and in the twentieth century with the automobile.[104] Take, for example, the famous poster of German physician Fritz Kahn (1888–1968), published in his book *The Life of Man* (1926), which depicts the heartbeat as a product of a pump. This demonstrates how, at the beginning of the twentieth century, the heart was conceived as a machine or motor because they were structurally akin. They shared some characteristics, such as the integration of energy, tubes, valves, and rhythms, as well as the need for electrical currents, oxygen, and feeding to be set in motion.

From the twentieth century on, artists and writers have equated the electrical and mechanical condition of heartbeats with those of lighting and mechanical machines. The connection between heart and light was based on the fact that both are electrical and mechanical energy generators with a rhythmic sound, and both can extend the liveliness of an individual: the heartbeat by preserving life and light prolonging its activities at night. In 1909, Filippo Tommaso Marinetti, in *The Founding and Manifesto of Futurism*, recounts having stayed up all night "beneath mosque lamps hanging from the ceiling" that "were illuminated by the imprisoned brilliance of an electric heart."[105] This analogy between dynamos and hearts is also found in the writing of the Russian doctor and writer Mikhail Bulgakov, particularly in his *Notes on a Cuff* (1920), in which a character declares, "In my heart a dynamo machine beats/ Ta, ta, tam."[106]

In *Multiplied Man and the Reign of the Machine* (1911), Marinetti proclaims, "Future man will reduce his heart to its purely distributive function. The heart has to become, in some way, a sort of stomach of the brain, which will methodically empty and fill so that the mind can go into action."[107] This literally implies a change of heart, in which futurists reject the idea of the heart as a "sensory receptor" vulnerable to love and lovesickness. On the contrary, they embrace the organ as an emotionless pumping machine whose function is the circulation of inextinguishable energy that keeps the body's mechanism active. Humans, according to Marinetti, were evolving into machines that had indefatigable mechanical hearts made of steel and pushed forward by speed. In his *To My Pegasus* (1908), the same year that Henry Ford opened the first assembly line to build an affordable

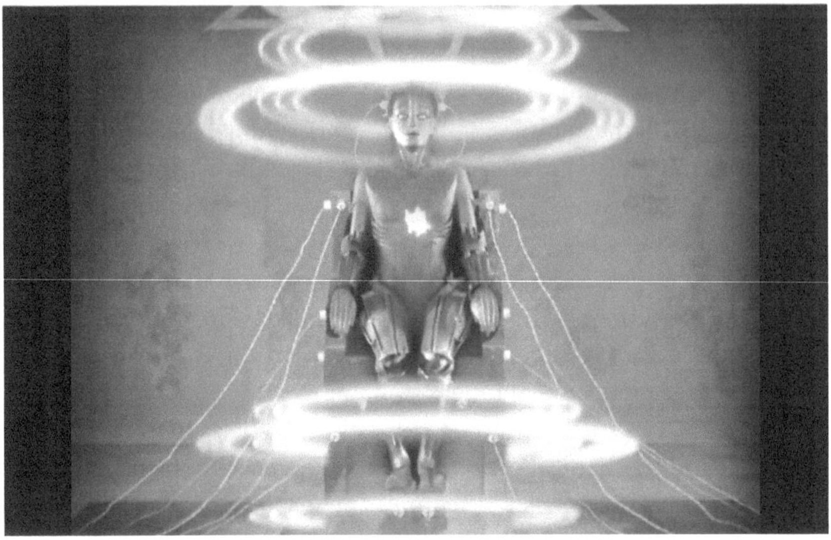

automobile, he describes how the thrill of velocity has the power to accelerate "bloody palpitations" that intoxicate and disconnect a car driver from earthly attachment in a sort of orgasmic sexual fusion between the human and the machine.[108] Marinetti's mechanistic and patriarchal vision prioritizes rationality over emotions, and reinforces the idea that the heart and its feelings needed to be managed and controlled as a mechanical system, while also diminishes humans relationships and the intricate web of relationships between our bodies and the natural world.

In the 1930s, the futurist artist and writer Benedetta Cappa, who was married to Marinetti, also followed a futurist masculinized rationality and linked pounding hearts, speed, and machines in some of her writings. In *Letter from Astra, Astra's Diary*, she narrates a romantic and mysterious

1.25

Still from Fritz Lang's silent film *Metropolis*, 1927.

dream with speedy flying machines that make the "motor hearts" of two lovers palpitate. Upon awakening, however, Astra is free of the anxieties of love energies and of "any attempt to rise up to the heights of optimism."[109] Clearly, both Cappa and Marinetti shared a fascination with new forms of transportation and deemed a fusion of flesh and metal as the ideal heart, putting industrialization and reason ahead of nature and emotion.

In Germany, Fritz Lang's silent sci-fi epic *Metropolis* (1927; figure 1.25) follows the Weimar avant-garde filmmakers' trend of equating the rhythms of the body with those of industrial objects and suggesting a tension between the organic and the technological.[110] In Lang's film, a "heart machine" is the city's power source; human Maria is the revolutionary counterpart of this heart machine, who claims that the heart should be a mediator between the head and the hands; and Maria's doppelgänger robot is a male-dominated figure, a pursuit to master nature, that has a luminescent pulsing heart that sends electrical currents throughout her mechanical body. This mechanized heart leads to the destruction of the heart machine and men, greed, and patriarchy.

Musing also in the connections of humans and machines, the same year that saw the release of *Metropolis*, the theater director Erwin Piscator, for his "theater machine," made an adaptation of Ernst Toller's *Hoppla, We're Alive!* For *Hoppla*, Piscator combined different technological elements, such as mechanical rhythms, slides, and film projections. As a symbol of humanity, heartbeats were produced by a machine and were the constant and most striking aspect of the scene. The scene opens with a radio operator announcing that a passenger of the first New York–Paris flight has had a heart attack and calls for a heart specialist. Then, loudspeakers reproduced the sound of heartbeats. This was followed by the projection of an image of an airplane over the ocean and X-ray images of a pumping heart. Piscator did not use actual heart rhythms as input; they were all produced mechanically.

A few decades later, artist Salvador Dalí presented *The Royal Heart* (1953), a late work created to commemorate the coronation of Queen Elizabeth II. The piece is a gold-crowned, heart-shaped jewel with rubies at the center throbbing like a beating heart. The pulsating effect was created by an electric synchron motor, which is normally used in clocks. The brooch is a shiny crafted representation of the beating heart using a mechanical procedure.

All these works are examples of how the heartbeat was represented in the first half of the twentieth century by mechanical means. However, the electrocardiograph and other technologies of the heart became widespread and accessible only after World War II, once physicians had improved it for commercialization by making it lighter, portable, and technically easier. Since then, the electrocardiograph has become a common tool in hospitals and doctor's offices around the world.[111] Artists quickly realized that cardiograms and heart sounds were not only a sign for detecting pathologies but could serve as a visual and audible representation of hearts and life. With its capability to extend the body outside the confines of the body, electrocardiograms had a great impact on the imagination of artists, who started to use electrocardiograms as an art material both for representing subjects and for making a sort of transmission of the self through technologies. Creative uses of cardiograms can be found from the late 1950s when various artists, such as Heinz Mack, Brian O'Doherty, or Mark Boyle and Joan Hills (all practitioners to be discussed in future chapters), used the electrical activity of the heart as an aesthetic media with very different connotations. Later developments in kinetics, lighting techniques, computer systems, and cybernetics (one of the earliest texts written by Norbert Wiener, Julian Bigelow, and Mexican physiologist Arturo Rosenblueth) in the mid-twentieth and early twenty-first centuries opened up other new and expansive possibilities. Artists such as Catherine Richards or Rafael Lozano-Hemmer used pulse sensing technologies to expand the interior of the body to exterior locations by means of different types of lights, screens, sensors, and projectors.

Undoubtedly, the electrocardiograph, the stethoscope, and the different models and machines made in the nineteenth century and developed further in the twentieth century opened up new creative possibilities to see, listen, characterize, and manipulate the heartbeat and the pulse. Artists embraced and expanded on these in the second half of the twentieth century, exploring the symbolic possibilities of such an apparently simple but significant rhythm. In the following chapters, we will see that after the 1960s, there were increasing and varied uses of the heartbeat as raw material, which will show the spatial, electrical, representational, regenerative, transplantable, and connective faculties of the heart.

2

SPACE

Your heart is a space with four chambers and two corridors or ventricles and occupies a specific space in your chest, slightly on the left side, behind your breastbone, and between your lungs. If you touch that part of your body above the chest, you can feel your heart beating. The pulse's expansive nature enables it to stretch not only across the body but far beyond.

During the second half of the twentieth century, scientific developments inextricably linked with the rise of military technologies and the space race enabled the electromagnetic signals of the heart to be seen and heard in other spaces, even beyond our planet. This chapter shows how the heartbeat is found not only in our body but also in the cosmos: in auroras, stars, and pulsars, as well as in the hearts of human and nonhuman cosmonauts who reached beyond Earth's limits. Cardiograms and recordings taken by biotelemetry were the first portraits of humans and nonhumans made beyond the stratosphere. The heartbeat was inserted into the spatial imaginary, both in the electrocardiograms of Laika and Neil Armstrong, as well as in the Golden Record, and in music (by Sun Ra or Pink Floyd) and movies inspired by science fiction. Inspired also by the moon landing and the discoveries in the universe, sound artists also started to deepen their perceptions of how the environment, human consciousness, and the body connected with the cosmos. Like our hearts, the universe has rhythmic variations or cosmic "heartbeat" patterns.[1] Thus, ontology is in the throbbing that connects us to other species' pounding hearts, as well as in the depths of the pulsating earth and the far reaches of the universe. A range of art production and strategies that connect heartbeats, art, and the cosmos are addressed in this chapter.

I analyze the projects of three artists, Pauline Oliveros, Alvin Lucier, and Richard Clar, who, being inspired by the satellite itself and the moon landing, set out to send body signals as a way of communicating with distant territories. I next dedicate time to Éliane Radigue's *Biogenesis*, a heartbeat-based composition that merges inner and outer space through "meditative modulations and quasi-cosmic vibrations"[2] that muses on the creation of life and its passage through constellations and generations. All artists in this chapter push past the limits not only of the internal and the external but also of the personal and the infinite.

Heartbeats, when connected to space, challenge the confines of the body, as well as the dualisms of terrestrial/extraterrestrial, inside/outside, whole/part, and many others. Artists have found them to be a symbolic, poetic, malleable, polysemic element that connects us equally not only to other human and nonhuman creatures but also to the universe, reminding us that everything, from stars to heartbeats, is throbbing, beating, and inexorably linked.

SPATIAL DATA: LAIKA, MOON LANDING

Picture a swirling burst of fluorescent green and violet light dancing across the vastness of a polar night sky as it chants eerie tones. Caused by energized solar particles interacting with Earth's magnetosphere, dazzling auroral displays create a fantasy of brilliant colors and vivid shapes that are forever changing. Witnesses to these aesthetic celestial phenomena have described their forms as oval or elongated arcs, with some pulsating appearing to have a "heartbeat" because of their flickering interaction with electromagnetic waves.[3] While their sounds are less known, they have been linked to the sound of heartbeats, drums, or chirping birds.[4] Auroras stretch hundreds of kilometers above our planet and are bright enough to be visible and audible from a satellite. Their powerful magnetic currents can interrupt radio signals and satellite communications when the atmosphere no longer stops the signal and, instead of bouncing it back to Earth, allows it to travel to outer space. Sixty years ago, an aureole was erased from a canine's heartbeat.

Laika's heartbeat is audible in *Voices of the Satellites!* (1958), a recording by the Smithsonian Institution compiling sounds of the first thirteen Soviet and American satellites launched into space. Ten strong and solid lub-dubs of the legendary dog's heart can be heard once it was cleaned of interference caused by the aurora.[5] The sounds could be recorded because the satellite in which Laika was traveling could be detected from an enormous distance, as it was itself equipped with a "heartbeat." The beeping sound, announcing the satellite's presence in space, was purposely designed to be tracked and heard by radio amateurs around the world to inspire admiration of the Soviet Union's technological achievements.

The Second World War and the subsequent Cold War were catalysts for significant advancements in electronics and medical technologies. These conflicts prompted the development of more efficient and compact devices, leading to the streamlining of technologies that were previously bulky and heavy. Medical technologies became more complex and were able to monitor the heart and breathing rates of people, even when they traveled beyond Earth's orbit. The possibility to record the effects of space flight on the human body opened up, but it was the dog Laika that led the way. Her vital signs were the first to be monitored in the stratosphere when, in 1957, she became the first animal cosmonaut to be launched into orbit (figure 2.1). The spiky lines of her heart rates were measured using a blood cuff implanted in her carotid artery connected to the monitoring system.[6] A recording of her heartbeat was transmitted back to Earth by biotelemetry, which was one of the technologies developed for space missions, first by the Soviets, followed by the United States.[7]

Laika's biological signals, which were intercepted by US intelligence, are the first telemetric recordings of a living creature's heartbeat in the exosphere: an outer space animal portrayal that is, in fact, a portrait of fear.[8] The black wavy lines tracking Laika's heartbeat spiked during the ascent from 103 to 260 beats per minute, tripling the dog's normal rate of around 80 to 130 beats per minute.[9] Finally, the pioneer's heart succumbed to stress and overheating.[10] Laika's heart chart portrays her brief survival beyond Earth's atmosphere, but, more importantly, it is a testament to

Publication No. 166 SECRET

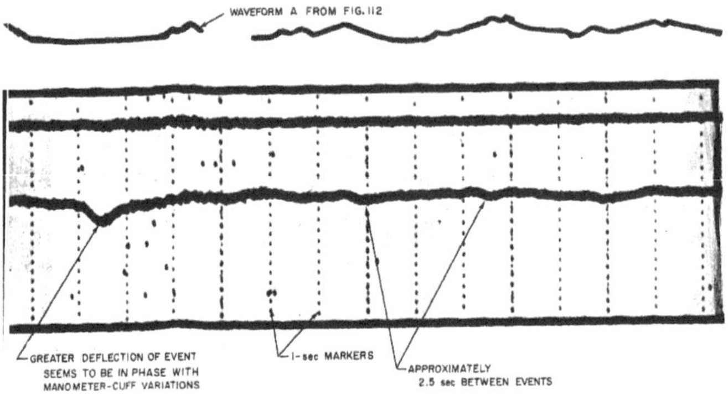

Fig. 119. *Sputnik-2* Telemetry Associated with Respiration

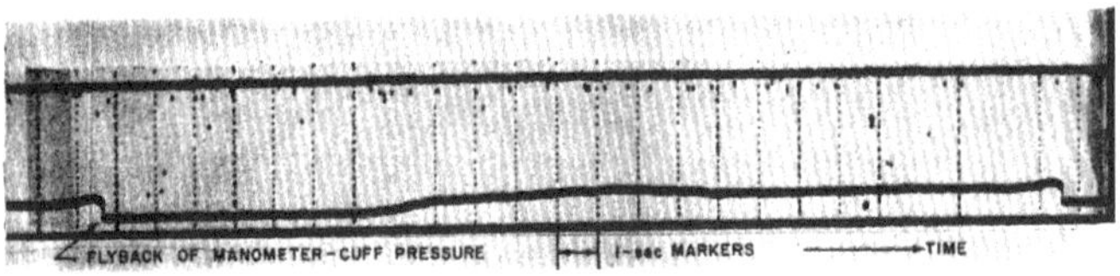

Fig. 120. *Sputnik-2* Telemetry Associated with Manometer-Cuff Pressure

Declassified and approved for release by NSA on 07-06-2006 pursuant to E.O. 12958, as amended

Page 86 SECRET

2.1

Laika's heartbeat intercepted from space in November 1957. Source: US National Security Agency.

how dogs have served in the human quests for medical knowledge, material resources, and the farthest reaches of outer space.

Many sentient beings, including mice, frogs, pigs, dogs, and horses, have been indispensable to advances in cardiac knowledge and the development of cardiac technologies. Subject to the imperative that preserves human biological life by the substitution of other biological entities, animals' bodies have been appropriated, regulated, and exploited as "models" of the human body by biomedicine. Indeed, space medicine had its origins in Nazis experiments on humans in concentration camps and was further developed by the Third Reich scientist Herbertus Strughold, who became head of animal experiments at NASA after leaving Germany in 1948.[11] The first hominid to launch into space was Ham in 1961. NASA records declare that Ham's heart rate remained within normal limits even when subjected to excessive forces and prolonged confinement. But the chimp returned distressed.

During space flights, heart rates become an indication of a terrestrial energy floating in another world, a form of communication in the face of uncertainty, a beat of hope. It is a form of existence where human life cannot be sustained, a life/nonlife form at a distance. Hundreds of kilometers from Earth, in a space which only deities, the dead, and aliens occupied previously, living beings are laboratory creatures or telecommunication messages. In this context, life and, therefore, mortality become something environmental since life is about living in a space capsule or a signal in telemetry. In space, as on Earth, the heartbeat is an index of vitality, but also of biopower and morbidity.

Animal astronauts were mostly employed to test one of the main challenges space physicians faced: managing the long-term effects of microgravity on astronauts' cardiovascular systems.[12] While on Earth, the heart pumps against gravity to push blood to our anatomical extremities; in outer space, the effect is the opposite. This results in an increase in the resting heart rate, a decrease in the size of the heart, and a reduction in blood pressure and arrythmias.[13] To counteract the effects of weightlessness in humans, scientists since the 1950s have conducted many studies and experiments to test physical and psychological reactions. Among the best-known was REST-flotation or the use of sensory deprivation tanks to create zero-gravity environments for the study of cardiovascular effects.

The heart of the Soviet cosmonaut Yuri Gagarin was the first to beat in space and safely return to Earth. His heart was monitored before launch and then during and after his flight. An EKG obtained the day before his flight on April 11, 1961, is on display at Moscow's Museum of Cosmonautics. The image, which serves as a portrait but also as a historical record, shows a stable wavy line, but his cardiac records hovered from 66 to 158 beats during the ascent. Surely, his heart also skipped a beat when he became the first human to contemplate our planet from a distance. However, just as his animal predecessors were monitored and recorded by technologies, Gagarin, too, contained in his spaceship, miles away from Earth, became both a subject and an object, a sign sent by telemetry.

The wonder of seeing our planet from afar in this science spectacle became a collective experience on July 20, 1969. The heart rates of millions of people around the globe accelerated when a previously inconceivable achievement was transmitted on television: the landing of Apollo 11 on the moon. Led by Neil Armstrong and Edwin "Buzz" Aldrin, the event, which is considered to be one of the most important technological achievements of the twentieth century, was transmitted live. The dramatic broadcast was carefully orchestrated by NASA and the US government as a counternarrative to the dominant news about the Vietnam War and political turmoil in the United States.[14]

Throughout NASA's "theatrics of space" operation, which was accompanied by a constant stream of clicks and beeps, the astronauts' heart rates were continuously recorded and announced by a NASA public officer (figure 2.2). Just a few minutes after launch, an update stated, "We have a report on the launch heart rates now from the flight surgeon. Commander Neil Armstrong's heart rate, 110. Command Module Pilot Michael Collins, 99. Lunar Module Pilot Buzz Aldrin, 88."[15] The data stream reflected the mission's pulse, its moments of challenge, calm, activity, and distress. During the lunar descent and ascent, for instance, an adrenaline rush was registered in Armstrong's heart rate, which increased to 150 beats per minute. In the mission's transcripts, other heart rates are mentioned: one of 96 beats per minute when Collins was exercising and a false alarm of 247 beats when Aldrin's sensors dried out.[16] In these isolated landscapes where living beings intrude, their cardiac energy, besides being a presence in

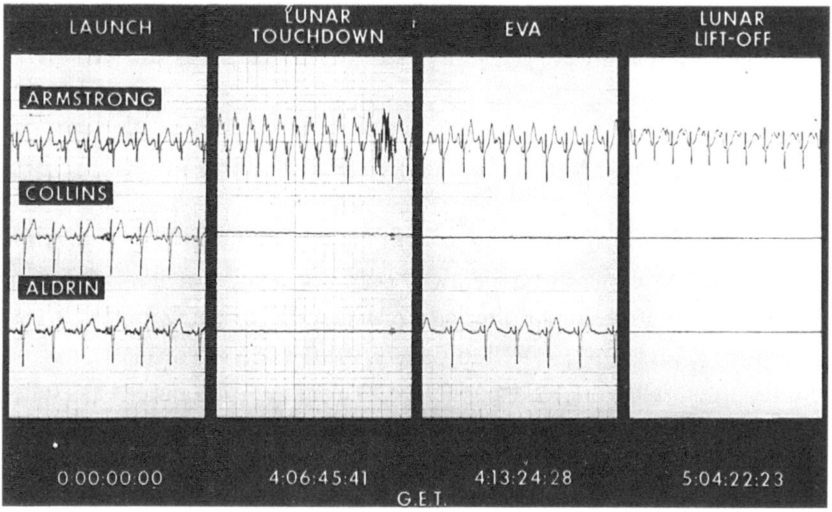

absence beyond Earth, was also subject and object, sensations and data, and an implosion of the natural subjected to the artificial, fact and fiction.

Artists' imaginations were sparked by the amazing visuals and beating sounds of outer space, and the grandeur and mystery of the universe of that previously unreachable space. The possibilities of seeing what's beyond the skies prompted artists to mirror themselves in space, explore the cosmos, grapple with existential themes, and combine scientific and technological aesthetics into their work to create spatial sounds or reach out into the universe. The beeping sounds recorded during space travels—and particularly heartbeats, those fluctuating sound waves that vary in volume and frequency—became a link between inner space and outer space, sound and emotions, the body and the cosmos. In the following sections, I show how heartbeats were interpolated into artistic

2.2

Typical EKG signal received during Apollo 11 mission. Source: NASA.

projects that were aimed to connect us to the universe in both physical and metaphysical ways.

GOLDEN RECORD: ANN DRUYAN, SCI-FI FILMS, PULSING STARS

In the vastness of the outer solar system, as a signal of life on Earth, a woman's heartbeat is still traveling aboard the Voyager spacecraft on two 12-inch gold-plated copper phonographs. Curated by astronomer Carl Sagan and a team from Cornell University, the Golden Record is a carefully crafted interstellar message conceived as a sign of our planet's existence (figure 2.3). Launched on its billion-year journey in 1977 in a spacecraft that is still sending a "heartbeat" signal of life from deep space today,[17] the albums aim to portray to any alien civilization the history, diversity, and evolution of terrestrial life. Through sounds and images of nature and the human world, it includes a selection of world music, a photo gallery of our planet's inhabitants, and an audio essay of natural and technological sounds titled "Sounds of Earth." Reflecting the reality that every sound in the universe begins with a vibration, the audio essay opens with the music of the spheres, followed by vibrating volcanoes, earthquakes, winds, rain, the croaking of a frog, the growls of a chimpanzee, and the barking of a dog. In a thirty-second clip that includes cardiac rhythms, we hear human footsteps, a steady heartbeat, and the laughter of what we can assume are early humans. These are followed by more sounds—fire, trains, tractors, buses, more machines. The compilation ends with the sounds of a kiss, a mother and child murmuring, an electroencephalogram (EEG) recording, and the steady beat of a pulsar.

Among all these sounds, the heartbeat has a special place on the Golden Record. It belongs to Ann Druyan, the creative director of the album, who at that time was in love. The idea of recording her own heartbeats and brainwaves occurred to her as she wondered if a highly advanced technology would be able to decipher human thoughts and feelings.[18] In a love letter she wrote to the two voyagers, Druyan relates how they were created and reaffirms their ontological mission and importance as time capsules:

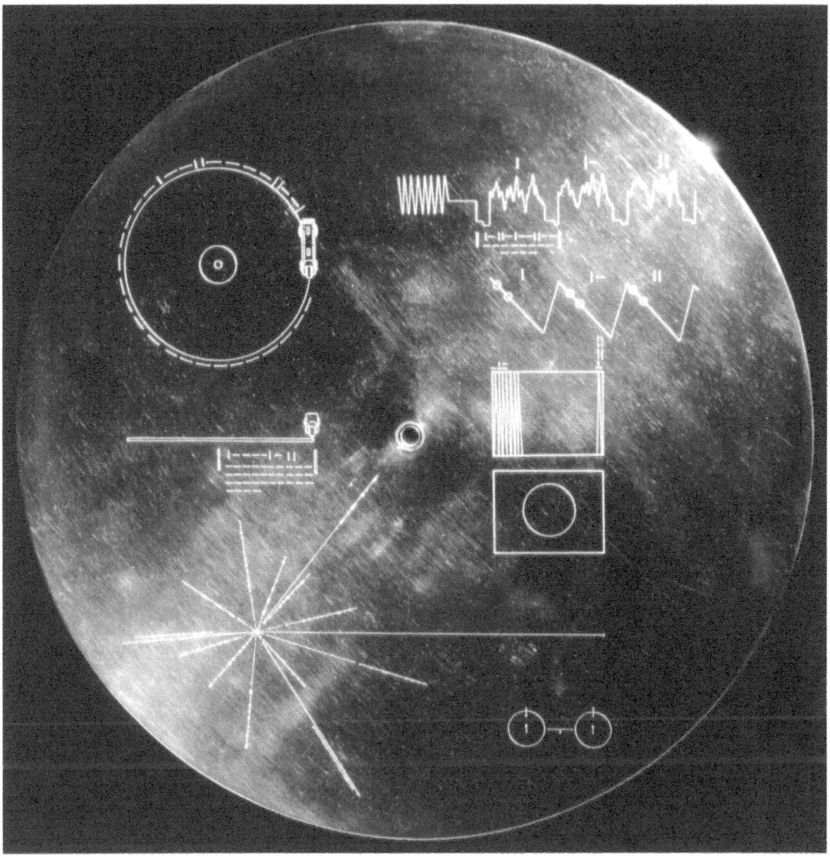

2.3

Voyager Golden Record, Sounds of Earth, Record Cover, 1977. NASA/JPL.

You, the farthest objects we have ever touched, now venture beyond that place where the sun's wind gives way to roaring interstellar gales; far, far away—and yet I feel close to you. My own life attained lift-off when you did. I was 27 years old when you were in the last stages of your assembly. Carl Sagan and I had known each other as friends and colleagues for a couple of years. We fell in love in 1977 while collaborating on the message that you carry. . . . Just days after Carl and I realized we would spend the rest of our lives together, I meditated for an hour while blindfolded, hooked up to every human monitoring device then known. The signals from my mind and heart—the most intimate message on the record—were translated into data and paired on the golden records with a strangely similar rasping, that of the most distant sound ever recorded back then: a pulsar. Now, I am seventy. Yet the feelings of that long-ago spring afternoon remain fresh and astounding. . . . Since you have exceeded your mission specifications in every other way . . . years from now . . . a heart at its greatest fullness, will sing on.[19]

The heartbeat on the Golden Record is a sign of life, but it is also a message of love of what one day might be the only remaining evidence of humanity. This takes the Golden Record beyond science to become a work of art: carefully selected and embedded with images, words, and sounds with symbolic connotations that open up new dimensions.

The Golden Record was originally created for an unknown extraterrestrial intelligence that might be interested or able to understand its contents as we do. This anthropocentric assumption is typical of the space age. Humans were anxious about nuclear annihilation and alien invasions, about space and the future of the planet, and the possibility of encountering "little green men" enthralled popular culture.

Science fiction movies evolved so that from the 1960s to the 1980s, movies appeared in which aliens were intelligent but heartless humanoids willing to exterminate humanity or as compassionate entities endowed with big palpitating hearts. The alien in *E.T.* (1982), the eponymous movie directed by Steven Spielberg, is a cute example of the latter, with his heart glowing in unison with his finger whenever he feels the emotions of humans. Yoda, from George Lucas's *Star Wars* (1977–), is another humanoid that uses the heartbeat as life guidance when he suggests, "Close your mouth and open your ears, you must. Then hear what your true heart is seeking."[20] In Gene Roddenberry's *Star Trek* (1966–), heartbeats are scanned on various occasions and symbolize the human side of the stoic Vulcan species, whose heartbeat rate is so fast that it resembles a vibration rather than a typical heartbeat. In contrast, Ridley Scott's *Alien* (1979) and James Cameron's *Aliens* (1986), two iconic movies that showcase fierce predators, use heartbeats as a dramatic tone that suggests the birth of a fearsome living organism and the harbinger of a life about to be lost. Furthermore, in Graham Baker's *Alien Nation* (1988), the Tenctonese language of aliens resembled a heartbeat graph coming from an electronic machine.

Fantasies of aliens that could occupy or communicate with humans through pulsations were not the exclusive province of science fiction but were also fostered by science itself. Take as an example an event that happened a decade before the Golden Record was released. When Jocelyn Bell, a trailblazer in a male-dominated field, discovered pulsars in 1967, she reasoned that the scintillating sequence of pulses was so precise that the signal must have been transmitted to Earth by an extraterrestrial intelligence. Bell even named the first pulsar LGM-1, referring to the stereotypical portrayal of aliens as little green men.[21] When a second pulsar was discovered, this theory was proved wrong. Today, we know that pulsars are very precise clocks that spin rapidly, emitting beautiful magnetized gamma rays at different rates. These electromagnetic waves have been converted into an audible signal that has been described as a heart rate.[22] The heartbeat-like waveform of pulsars has inspired some works of art. The iconic cover of Joy Division's debut album, *Unknown Pleasures* (1979), is in fact the signal of pulsar LGM-1, also known as CP 1919.[23]

Exploring the vast reaches of space, astronomers, through their telescopes, have detected the pulsating "heartbeats" of distant stars and galaxies, including those of pulsars, pulsating auroras, miniature black holes, gas clouds, Delta Sculti stars, and "heartbeat stars" (so-called because the tracing of their brightness draws an oscillating wave that resembles an EKG; figure 2.4).[24] The rhythmic pulsations are sound waves originating from inside each star, producing different frequencies. Moreover, human heartbeats seem interconnected to the universe as well: recent studies have determined that changes in the solar and geomagnetic environment—such as cosmic rays, solar radio flux, and Schumann resonance power—can have an impact on human heart rate variability and parasympathetic activity.[25] These connections between our hearts

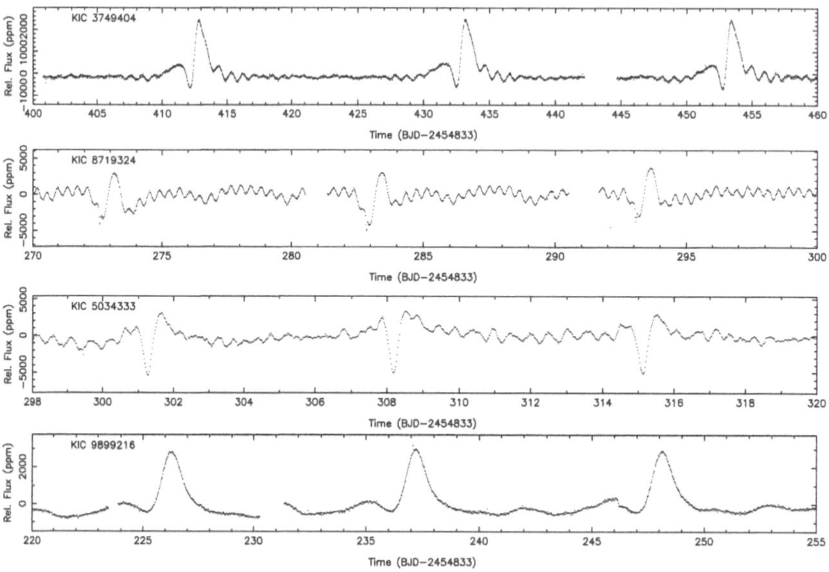

and space are being enhanced today as astronauts conduct research on bio-printing hearts in 3D, which mimic the beautiful form and textural plasticity of the real organ. By printing living tissue in space's unique microgravity environment, they look to investigate the heart's aging process (which is twenty times faster in space) and create artificial hearts for future transplants.[26]

Heartbeats, as electromagnetic waves, are microcosms of the larger rhythms and pulsations that occur on Earth and throughout the cosmos. In the mesmerizing tapestry of pulsating rhythms, the heartbeats on the Golden Record, the heartbeats observed by astronomers in space, and the heartbeats of aliens in movies converge, uniting worlds and bridging the vastness of the cosmos.

2.4

Light curves of heartbeat stars.
Source: Nasa Ames Research Center.

BEAT MUSIC: SUN RA, MILFORD GRAVES, PINK FLOYD, KRAFTWERK

The contact of humans with outer space had a strong impact on the general public through popular culture and on musicians who started to create works that related heartbeats with the cosmos and the future. Inspired by physical incursions and technological developments from the late 1950s and 1960s, and the increasing production of science fiction imagery, Space Age Pop emerged, incorporating unearthly sounds interwoven with metaphors of space and alienation. Continuously broadcast across radio and television, Sputnik's beep became the sound of outer space, the "life-pulse of nascent otherworldly soundscapes."[27] Russell Garcia's *Fantastica: Music from Outer Space* (1959), for instance, starts with a countdown announcing humans launching into outer space, followed by beeping sounds.

Combining ideas of space, science fiction, Blackness, and technology, Afrofuturism featured a cosmic vision, which, in contrast to the white-led space project and imagery, sent African Americans to space.[28] In the 1950s, Sun Ra (figure 2.5), the Afrofuturist pioneer who had a profound impact on rock, pop, and experimental music, began to dress in futuristic Egyptian-like clothes and claimed to be an alien from Saturn.[29] He was inspired by sci-fi rhythms and developments in technology, such as the newly invented Minimoog synthesizers. While his music didn't incorporate actual heartbeats, *Space Is the Place* (1972) and his subsequent albums incorporate electric keyboards and steady beats. His "music pulsing like a living heartbeat" has a vibrant and cosmic vitality that transports his audiences to deep space.[30]

Milford Graves, the Jamaica-born iconic 1960s cosmo-spiritual free jazz musician and artist, once said, "Throw away your metronome and listen to your heart."[31] Inspired by traditional African ritual dances and drumming, Latin/Afro-Cuban rhythms, Chinese medicine, Zen philosophy, and the study of human biology, he studied the cosmic relationship between the drumbeat, the heartbeat, and the universe. During his lifetime, he understood the musical instrument that he mastered in an expansive and holistic way. He made music notations and visualizations of the heartbeat, in which he later found an actual form of healing hearts through vibrations and cells.[32]

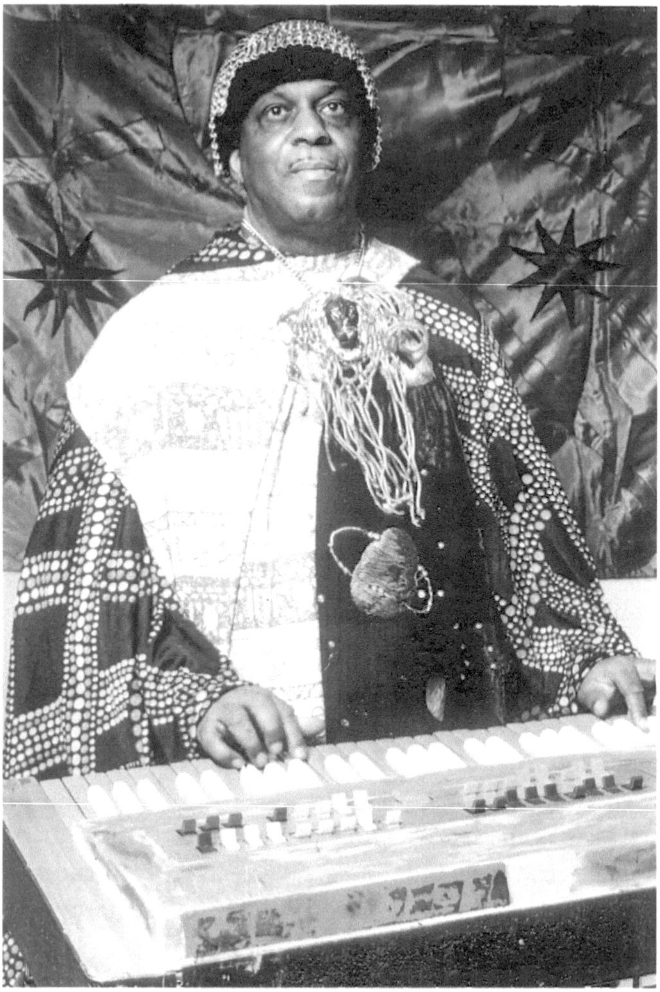

2.5

American jazz musician Sun Ra promoting the 1973 reissue of his album *The Magic City* through Impulse! Records and ABC/Dunhill Records.

Bombarded by images of space travel and the future in film, literature, and advertising, space and fantasy became the pulse of pop and rock music. Space noises and heartbeats became a recurring presence, opening up a new repertoire of sounds to represent life on Earth and in the cosmos, as well as to conjure up the influence of drugs, human experience, the passage of time, and the interactions between humans and machines.

Over this period, the advancements in the Moog, ARP, and VCS3 synthesizers and recording technologies allowed psychedelic rock artists to experiment with "sound effects—such as explosions, heart beats, and the sound of breathing" to create what they called "spacey music."[33] This was music that channeled interest in space through spatial and psychedelic sounds that gave the sensation of being in the cosmos and of being high or "spaced out." The music of both David Bowie and Pink Floyd at this time was so related to space that David Bowie's "Space Oddity," which tells the story of the fictional astronaut Major Tom, was played during the British broadcast of the 1969 moon landing, while the BBC commissioned Pink Floyd live jamming "Moonhead," with its combination of atmospheric sounds and heartbeat-like drumbeats. Other songs like David Bowie's "Five Years," which announces an apocalyptic disaster, and Pink Floyd's *The Dark Side of the Moon* (1973) which is an inner journey through several stages of life, begin or end with a bass drum heartbeat that gradually increases in volume and serves to set an introspective, suspenseful tone.

By mixing corporeal and unearthly sounds, the space music of Bowie, Pink Floyd, and other progressive rock musicians like Hawkwind or Gong created space journeys that led to impassioned existential inner journeys. Heartbeats were the pulse of life that mixed with psychedelic noises to channel the interest in space while also being a vehicle of emotional resonance and a symbol that evoked the passage of time, vulnerability, and human experience.

The increased production of sci-fi movies in the late 1970s and '80s, particularly *Star Wars*, served to popularize the genre on a never-before-seen scale, giving rise to space-related pop music.[34] Michael Jackson, the "king of pop," had an interest in space that not only led to his participation as the lead in George Lucas's and Francis Ford Coppola's *Captain EO* (1986) but also to his famed moonwalk. Considered a play on gravity, the moonwalk was marvelously performed by the singer in his video for

"Smooth Criminal," a song that begins with a recording of the singer's own heartbeat. During the same decade, pop songs in honor of Laika were written in different parts of the world, partly as a result of Lasse Hallström's film *My Life as a Dog* (1985), which features the canine astronaut. One of the first of these songs is by the Spanish band Mecano. This lament over human ambition and the sacrifice of a dog opens with a heartbeat-like drumbeat.

Techno and electropop, prominent in the 1980s, were also influenced by sci-fi and the advancements in technology, which permeated songs with strong heartbeats and drumbeats that interfaced with the body, machines, and cyberspace. In Detroit, the cradle of techno, pioneer musicians Juan Atkins, Derrick May, and Kevin Saunderson, inspired by Afrofuturistic music, created a novel sound of throbbing bass and looped beats of 120 to 150 bpm (beats per minute) that can accelerate your heartbeat.

Inspired by a variety of sources, including James Brown's funk music, Kraftwerk's electropop blends synthesizer sounds, drum machines rhythmic patterns, computerized beeps, mechanical noises, and vocoders to give their music a science fiction-esque tone. In Kraftwerk's *Man Machine* (1978) and *Computer World* (1981), booming "soundscapes" have the unusual power of allowing you to hear your heartbeat "loud like . . . in an anechoic chamber."[35] Most importantly, "Tour de France" (1983) and "Elektro Kardiogramm" (2003; figure 2.6) are two songs that incorporate heartbeat recordings as raw material alongside breathing and mechanical sounds. Indeed, the official video of "Elektro Kardiogramm" begins with scenes of outer space, followed by breaths and heartbeats. The video depicts wavy EKG lines behind Kraftwerk lookalike robots dressed in neon-lit-wet suits while "minimum/maximum beats per minute" can be heard.

In addition to creating an emotional and immersive experience, heartbeats added a sense of rhythm that made music danceable or, in the case of space rock, imbued it with an otherworldly hypnotic tone. Heartbeats and heart beeps were signs that represented life in humans and nonhumans, and they served to create a futuristic atmosphere that reflected the cultural fascination with technology, space, and the rise of the digital age.

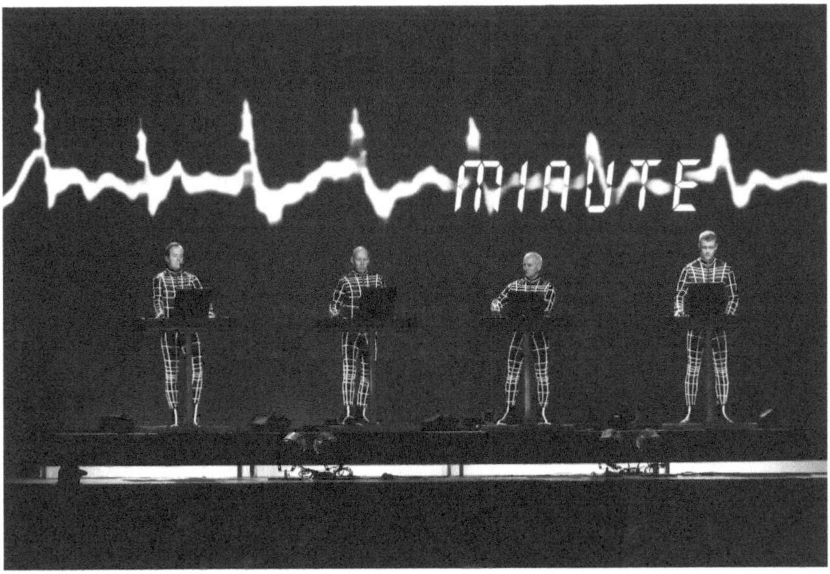

MOON BOUNCES: PAULINE OLIVEROS, ALVIN LUCIER

Close your eyes and listen to the external and internal noises around you. The beat of your heart, the pulsing of water, the sound of air hitting leaves. Then try to connect what you hear to the sounds of the universe. Linking all these noises from the inside, the outside, and beyond, one not only wonders if we are alone in the cosmos but also if our bodies are interconnected with the universe. In the late 1960s, Pauline Oliveros, one of the most influential electronic music composers and a trailblazing queer feminist, developed her intimate participatory art with profoundly embodied listening. Her pieces were created with acoustic and electronic instruments (figure 2.7), aural spaces, and an array of noises. Influenced

2.6

Kraftwerk playing "Elektro Kardiogramm" in a concert in Kyiv, Ukraine, September 27, 2008. Photo © Sérge Illin.

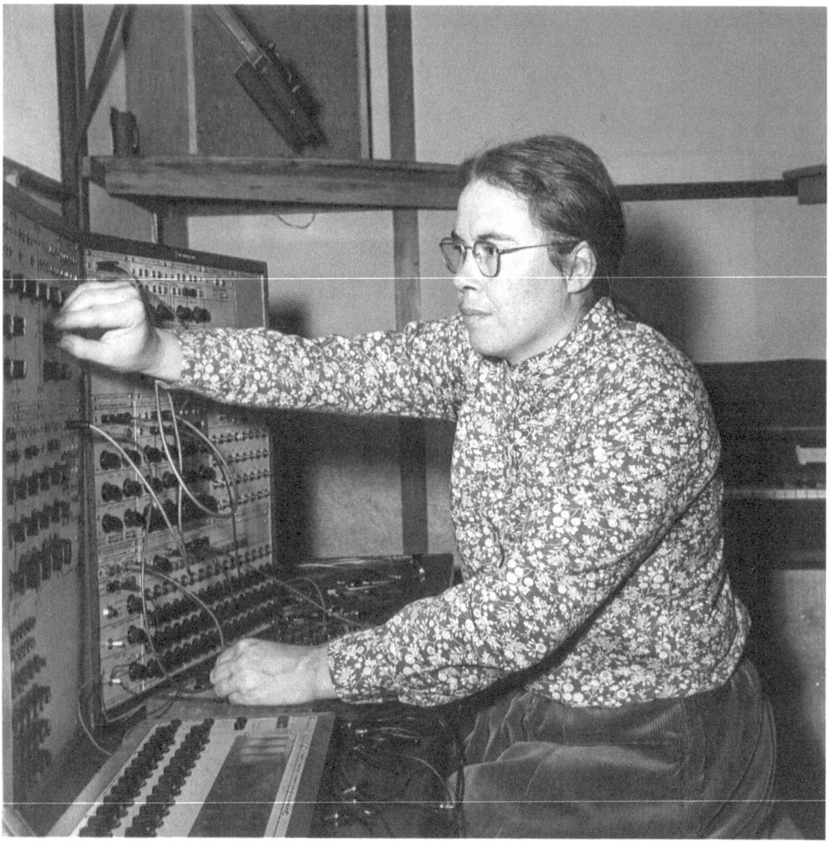

2.7

Pauline Oliveros with a synthesizer, 1966. Courtesy of the Center for Contemporary Music Archives, Mills College.

by Zen Buddhism, space travel, and technological innovations, she began to concentrate her auditive practice on listening to earthly and unearthly noises, the inner body, the environment, and beyond.

Listening to the body, and particularly to the heartbeat, was a practice Oliveros began in the late 1960s while pursuing her Deep Listening, which she described as the total attentiveness to the vastness of sounds in "the whole space/time continuum" to expand consciousness.[36] Oliveros's interest in body sounds was part of a wave of somatic advocates who, in this period, began to pay more attention to the "intuitive body" than to the rational mind. The trend was fueled by a wide range of practices that arrived in the United States at that time, including yoga, tai chi, chi kung, and German gymnastics. All of them were important to Oliveros, particularly through the work of her teachers, Elaine Summers and Al Huang. From Summers, Oliveros learned the "kinetic awareness" method, which put attention on the signals of her body and its almost imperceptible movements. With Al Huang, Oliveros studied tai chi and learned that *chi* (or *qi*), the vital energy of the universe, flows through all living things.[37] Oliveros's composition *Energy Changes*, published in her *Sonic Meditations* (1971) and dedicated to Summers, invites participants to listen to the environment as a drone sound, establishing a mental contact between the external sounds of the environment and the internal sounds of their own bodies, such as heartbeats.[38]

-XIII-

Energy Changes

(For Elaine Summers' movement meditation, *Energy Changes*)

Listen to the environment as a drone. Establish contact mentally with all of the continuous external sounds and include all of your own continuous internal sounds, such as blood pressure, heart beat and nervous system. When you feel prepared, or when you are triggered by a random or intermittent sound from the external or internal environment, make any sound you like in one breath, or a cycle of like sounds. When a sound of a cycle of sounds is completed re-establish mental connection with the drone, which you first established before making another sound or cycle of sounds.[39]

Energy Changes encourages people to pay attention to things that are normally ignored, such as heartbeats and blood flow, in order to deepen their connections with themselves and the environment. In Oliveros's work, listening to heartbeats is a call to self-awareness and forming connections between the mind and the body, and between the inner space and outer space. Oliveros's practice implied a spatiality that began with the body and the immediate environment, and expanded to include the universe. Her interventions in space, in contrast to those directed by space travel, were nonpatriarchal, noninvasive, and creative of symbolic linkages with the universe through music. This is evident in one of Oliveros's first heartbeat awareness works, *Indefinite Integral of Psi Star Psi D Tau Equals One* (1969), staged in a gymnasium a few months before the moon landing in February 1969. The two-hour interoceptive performance involved heartbeats and the creation of a constellation of the night sky on the floor, using flashing lights as stars. In her instruction score, Oliveros wrote that there were many performers, some with cellos, bassoons, or shakuhachi, others who were reader-singers. Heartbeats, music, and meditation were intertwined, particularly in the cardiac rhythm of two improvising artists playing the shakuhachi, a Japanese bamboo flute that was amplified and modulated by two electronic operators.[40] In the gym, making a constellation, all performers and the audience were located according to their date of birth, forming a constellation. They were instructed to meditate and use extrasensory perception (ESP) to try to produce silence and sounds of specific pitches within their own constellations.[41]

Oliveros's choice of the traditional Japanese shakuhachi as the instrument for the performance is significant. The end-blown flute has a long history intertwined with Zen Buddhism and serves as a useful tool for meditation because of its inhalation/exhalation rhythm, which has a positive effect on the heart rate.[42] The heartbeat, that fundamental rhythm that is constantly present but rarely noticed that is found in our bodies, as well as in the pulsating stars, is an element that connects to our inner life and the life around us. In *Indefinite Integral of Psi Star Psi D Tau Equals One*, the sound of heartbeats combined with that of the wind instrument and the use of ESP was intended to set an immersive and mediative guiding rhythm that not only inspired interpersonal connections and an awareness of the body and its vital signs but also of the position of the body in the universe.

Oliveros's approach to improvisation and composition was guided by what she called the sonosphere. This is "the sonorous or sonic envelope or the earth created by all vibrations set in motion by natural or technological forces."[43] According to Oliveros, these audible and inaudible vibrations, which involve heartbeats and the sounds of the stars, are processed consciously or unconsciously, and they could open up a new way of understanding life and the universe. As noted by Douglas Kahn, Oliveros's sonosphere "resonates among personal and interpersonal, musical, earth and cosmological scales informed by physics and metaphysics."[44]

Space, the sounds of the cosmos, and life beyond Earth appear in many of Oliveros's compositions and writings. In "Quantum Improvisation: The Cybernetic Presence" (1999), she suggested a new musical paradigm developed by technologies, in which sounds could be found at a cosmological scale and could set "interdimensional galactic improvisations with yet unknown beings"[45] and "the ability to detect locations from light years away."[46] In that same text, she wrote,

> What would a spatial melody sound like—a pitch beginning on Saturn moving to Aldeberon [*sic*] to Sirius to Earth? . . . Who would be playing this tune? Who would be listening and where? Melody across space stretched out and also happening everywhere simultaneously. Space is the place—I hear you Sun Ra![47]

By referring to Sun Ra, Oliveros endorses his belief that humanity is part of something greater, vaster, and cosmic in nature and that music as sound wave, in all its forms, is a medium for connecting to the universe and its cosmic pulsing rhythm. Inspired by space travel, the sci-fi imagination, and technology, Oliveros connected the vibrations of the body to the moon. Watching the 1969 lunar landing kindled her passion for the potential of communication far beyond Earth: "I saw the astronauts landing on the moon," Oliveros said, "I thought, the moon reflects light, why couldn't it reflect sound? Why couldn't I send sound to the moon and get the echo to come back?"[48] This event inspired *Echoes from the Moon* (1987), in which she transformed participants into "vocal astronauts" by sending their voices to the moon and hearing the echo back on Earth. Scott

Gresham-Lancaster, who assisted her in putting the piece together, put Oliveros in touch with amateur radio operator David Olean, who was one of the first to perform a moon bounce using a Yagi antenna. Oliveros sent her first lunar hello from Olean's studio; then she played a duo with the moon, making her accordion, tin whistle, and conch shells reverberate.

The sensation of touching the moon with her voice was so compelling that Oliveros repeated the experience at California State University a decade later, during the lunar eclipse of September 1996. During this event, hundreds of people of all ages happily lined up to hear their utterances "processed" by the moon in two and a half seconds.[49] The composer recalled that the experience of "touching the moon" with their voices after they traveled hundreds of thousands of kilometers away from our planet was such a great kick that the audience cheered each time their echoes returned.[50]

Our voices, like our heartbeats (which are essentially the voices of our hearts), can be a form of self-recognition because "hearing one-self speak," as noted by Jacques Derrida, is a form of auto-affection that is fundamentally hetero-affection.[51] "The voice is the being which is present to itself in the form of universality, as consciousness; the voice is consciousness."[52] Being conscious of our heartbeats implies being conscious of our vulnerability, transitivity, and finiteness. A voice or a heartbeat on the moon is a passage to the infinity of interiority, an exit from the world to another world, an expansion of consciousness through reverberations. A type of energy that, like heartbeats and pulses, generates intimate contact and transcends us.

One of Oliveros's final works that incorporated heart rhythms was the virtual reality interactive performance *Rotating Brains/Beating Heart* (2010–2012). The piece explored collaborative relationships between the mind and the heart, real life and virtual life, and inner space and outer space using sensing technologies. It was created in collaboration with Tina Pearson, the Avatar Orchestra, and Stelarc (an artist who has also amplified his heartbeats in many of his performances).[53] In the virtual space, giant hearts and brains were like floating planets in the sky, emitting the sounds of heartbeats, breathing, circulation systems, and mechanical devices every time an avatar of the artists touched them in real time. Using a virtual mixer in real life, Oliveros combined heartbeat sounds, brain waves, drones from her *Heart of Tones* (1999) piece, and words spoken by Stelarc. The work, which is mostly a game, brought together performers who were geographically distant but connected in virtual space. The game's symphony linked the brain and heart, inner space and outer space. The piece conveys the idea that the sounds that emanate from every heart and brain are each a universe and are each interconnected.

Pauline Oliveros died in 2016, leaving behind a significant legacy that advanced the radical position of turning inward to heed the sounds of the body and the exterior world. Although she never sent a cardiac voice to the moon through moon bouncing, years later, another artist did so, using her moon-related work as a precedent. In October 2018, two years after Oliveros's death, her good friend Alvin Lucier, known for his groundbreaking *Music for Solo Performer* (1965) using brainwaves, produced the heartbeat-based *Moon Bounce* (2018). The work recalls Oliveros's *Echoes from the Moon* and *Indefinite Integral of Psi Star Psi D Tau Equals One*. In their pieces, both Oliveros and Lucier blend heartbeats and the environment like early nature-based description of pulses did (see chapter 1). *Moon Bounce* was presented as part of the public program supporting Tomás Saraceno's exhibition *On Air* at the Palais de Tokyo in Paris. Saraceno's show was a choreography of human and nonhuman voices that revealed the intertwined and fragile rhythms that link earthly and cosmic worlds at a time of rising alarm over climate change.[54]

For *Moon Bounce*, Lucier picked up his heartbeat using a special sensor and routed the signal to a *qin*.[55] Lucier's choice of a qin was a symbolic one because, traditionally, in ancient China, the instrument was believed to link men and the cosmos. Its shape was seen to represent the sky and the Earth, and some believe it has healing properties that affect the human heart rates. The combination of low lighting with the subtle and ethereal sound of the artist's cardiac thump, filtered through the ancient Chinese instrument, created an intimate and relaxing setting. It is claimed that the performance of any qin signifies the union of Earth, humans, and the universe.[56] And Lucier's performance certainly interconnected all three. Using an Earth–Moon–Earth circuit, Lucier's heart waves were sent from Paris to the moon by a series of radio telescopes stationed in Italy, Switzerland, and the Netherlands. The two-way journey of the heartbeat took two and a half seconds to return to Earth with a new sonic character. This change occurred because the irregularities of the surface of the moon, from which the signal is reflected, altered the sounds and syncopations in Lucier's heartbeat.

As Lucier explained in an interview, he was not so interested in the body but in creating pieces of music: "The whole point of the piece is bouncing sound off of the moon. It will be hitting different spots as the moon spins around, hitting each location with a huge delay. The sound should change, and I want to let that happen."[57] *Moon Bounce* was a single composition made in collaboration with the moon; it is a heartbeat lunar song or, more precisely, poetic cosmic bodily music.

Another take on a heartbeat-moon bouncing is *Giant Step* (2015–2019) by American artist Richard Clar, whose work for decades has focused on the aerospace industry. For the first part of this work, in 2015, the artist translated Armstrong's famous words, "That's one small step for man, one giant leap for mankind," into Morse code and sent them to the moon as a pulsing green beam of light from the Côte d' Azur Observatory's Lunar Laser-Ranging Station. Following that event, he decided to turn Armstrong's EKG into music with the help of data scientist Ryan Compton. He converted the EKG into sound, and jazz musician Robert Mirando made a composition inspired by the EKG and the Morse code of Armstrong's famous words. On the fiftieth anniversary of the moon landing, Armstrong's heartbeat and words were sent back to the moon at the exact time the astronaut first stepped onto the lunar surface. The transmission was sent via satellites located at the Observatoire de la Côte d' Azur, France, and at the Dwingeloo Radio Observatory in the Netherlands. By symbolically sending Armstrong's heartbeat out into space, Clar paid tribute and celebrated Armstrong's first steps on the moon.[58] On this occasion, though, Armstrong's presence in space was less intrusive, and his heart did not have to withstand the physical and emotional challenges of being in outer space.

Space exploration has allowed us to understand our universe better. In the same way, art helps us understand, and often break, the models and assumptions of our reality. Sending heart voices to the moon is a form of communication through a universal language based on vibrations. Oliveros's influential concept of the sonosphere extends our understanding of the body, the self, and the natural world by locating us as part of the universe, which is itself infused with auroras, pulsars, and stars. The hearing of heartbeats in Oliveros's works serves as a bridge to connect us to other pulsations in the universe and expands individual and universal consciousness.

UNIVERSE'S VIBRATING SYMPHONY: ÉLIANE RADIGUE

You're listening to a track that starts with a gentle, spacey rumble; then, it gradually flows into soothing beats with subtle variations, heralding the beginning of a new life. This is the work of the French avant-garde composer Éliane Radigue. She was born in 1931 in Paris and became one of the central figures of experimental electronic music. When she learned that she was to become a grandmother, she expressed her joy by using her future granddaughter's heartbeat to create one of her most significant and personal pieces of music. *Biogenesis* (1973) is a 21-minute synthesizer-based composition that was first presented in 1974 at The Kitchen, a New York City not-for-profit space where many composers in the 1970s presented their new ideas (figure 2.8). The composition begins with a sound that transports you to the vastness of space and then progressively blends with hypnotic pulse rhythms and occasional beeping sounds to create a subtle but powerful musical experience. The heartbeat sounds in Radigue's composition, as on the Golden Record are a love message. The dedication of the piece reads, "To my daughter, Anne Person, for the child she carries and will carry for a long time."[59]

Biogenesis, which Radigue considers a purely emotional and experiential piece, evokes *omnis cellula e cellula*,[60] or in the artist's words, "the genesis of a biological process, the development of tiny cells that led to beings such as us."[61] Radigue described the structure of this work to me in an email:

The beginning of the piece is the heart beating of the father, then of the mother (my daughter), going to her belly to catch the beating of the heart of the baby, which is at the end of the piece, combined with a lot of internal noises of the body. Of course, all these basic elements are worked out carefully after with my synthesizer. This was a way to celebrate the birth of my first grandchild, who

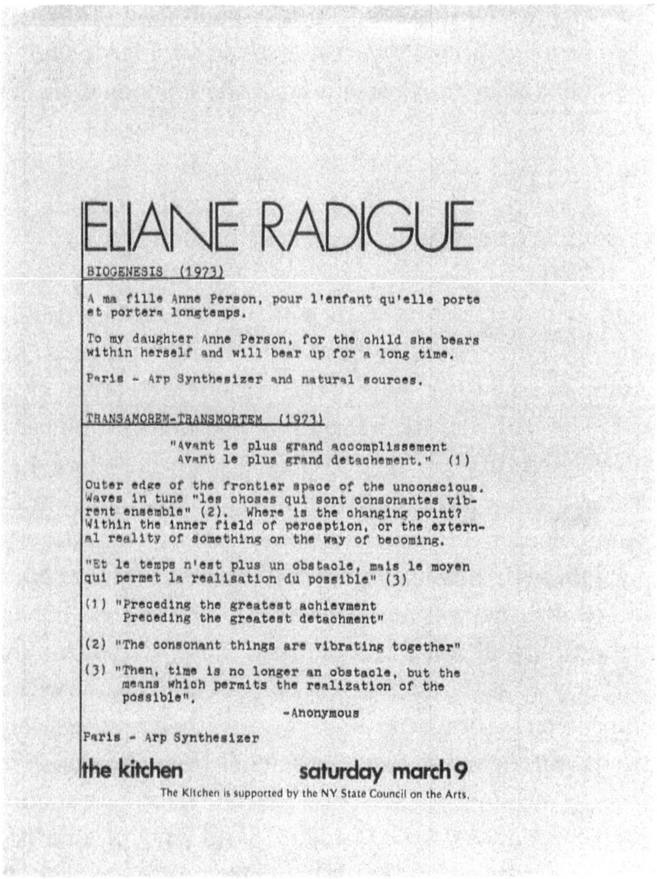

ELIANE RADIGUE

BIOGENESIS (1973)

A ma fille Anne Person, pour l'enfant qu'elle porte et portera longtemps.

To my daughter Anne Person, for the child she bears within herself and will bear up for a long time.

Paris - Arp Synthesizer and natural sources.

TRANSAMOREM-TRANSMORTEM (1973)

"Avant le plus grand accomplissement
Avant le plus grand detachement." (1)

Outer edge of the frontier space of the unconscious. Waves in tune "les choses qui sont consonantes vibrent ensemble" (2). Where is the changing point? Within the inner field of perception, or the external reality of something on the way of becoming.

"Et le temps n'est plus un obstacle, mais le moyen qui permet la realisation du possible" (3)

(1) "Preceding the greatest achievment
Preceding the greatest detachment"

(2) "The consonant things are vibrating together"

(3) "Then, time is no longer an obstacle, but the means which permits the realization of the possible".

-Anonymous

Paris - Arp Synthesizer

the kitchen **saturday march 9**

The Kitchen is supported by the NY State Council on the Arts.

2.8

Éliane Radigue, flyer for performance of *Biogenesis* (1973) and *Transamorem-Transmortem* (1973) at The Kitchen, March 9, 1974. Courtesy of the artist and The Kitchen.

happened to be a girl. She is 46 now; she has three boys from 21 to 15 years old. The dedication is a kind of homage to the transmission throughout ages and generations.[62]

The source of the sounds used in Radigue's *Biogenesis* is difficult to identify, and they are not easy to recognize as heartbeats. While some of the beating sounds seem to parallel the sound of planetary signals retrieved at great expense from space, the instruments she used to record these cardiac rhythms were all relatively modest:

1. A Sennheiser quite good microphone.
2. A medical stethoscope.
3. A tiny bottle that I find by chance in a store for children. I cut the bottom, which was of the size of the microphone, and I used the top in one of the branches of the stethoscope, and the second branch in my ear to listen to the recording. All that maintain[ed] by small cords. Very primitive.

All these instruments are now in a box, which I gave to my granddaughter.[63]

Radigue's attraction to heartbeats can be traced to her first encounter with *musique concrète* in her early twenties. Hearing the puff of steam engines and skidding wheels in Pierre Schaeffer's *Étude aux chemins de fer* (1948) on the radio was a eureka moment that helped her realize that all sounds could be music. From 1955 to 1957, Radigue studied electroacoustic techniques with Schaeffer and Pierre Henry at the Studio d'Essai de la Radiodiffusion Nationale, in Paris. There she began to experiment with feedback and tape manipulation, techniques that prepared the ground for her heartbeat composition and her subsequent work: "taking real sounds and manipulating them by cutting, splicing, editing, slowing down and so on."[64]

After focusing her attention on their children for nearly a decade, in 1963, Radigue moved to the United States for six months with her husband, the French-born American artist Arman. In New York City, she met like-minded and inspiring artists working with subaudible and minimalist sounds, including James Tenney, John Cage, David Tudor, and Philip Glass. Minimalist composers like Steve Reich and La Monte Young were inspired by the vastness and emptiness of space, which often translated into the repetitive and expansive nature of their musical compositions. On her return to Paris, Radigue found that her new work was met with disinterest by Henry and Schaeffer, who detected in it an unwelcome shift toward American experimental music.[65] "I'm from a generation where the machos were overwhelmingly prevalent, and the only way I found to deal with that was to ignore them. I didn't feel like fighting at all."[66] Feeling isolated, she decided to return regularly to the United States to continue experimenting with her compositions.

During a residency at New York University in the 1970s, she discovered the instrument to which she would devote herself for nearly forty years: an APR synthesizer. With this instrument, which she used without a keyboard, she created her signature style of spacey, delicate, pulsing, low, and slow-evolving forms with sustained tones, as featured in *Biogenesis* and many other of her greatest pieces. Radigue outlines her extensive and lasting interest in music, pulsations, and life transmission in her 2009 text "The Mysterious Power of the Infinitesimal":

> In the beginning, there was the air's powerful breath, violent intimidating tornadoes, deep dark waves emerging in long pulsations from cracks in the earth, joined with shooting fire in a flaming crackling. Surging water, waves streaming into shimmering droplets . . .
>
> Was it already sound when no ear was tuned to this particular register of the wave spectrum in this immense vibrating symphony of the universe? Was there any sound if no ear was there to hear it? . . .
>
> Then came the electronic Fairy; through the power of magnetic, analog and digital capture, breath, pulsations, beating and murmurs can now be defined directly in their own spectrum, and thus reveal another dimension of sound—within sound . . .
>
> Another story was beginning. A story where breath, pulsations, beating, murmurs and above all the natural production of these marvelous, delicate and subtle harmonics could be deployed in a differently organized manner.[67]

In *Biogenesis*, Radigue combined her electronic vocabulary of sustained sounds, pulsations, and beatings with the internal rhythms of the body, following her mood. As the critic Geeta Dayal has noted, her approach to music is "less about controlling sounds and more about surrendering to them."[68] *Biogenesis*'s introspective tone, lack of metrical timing, resonant and low beats, and sustained tones foreshadow Radigue's interest in Tibetan Buddhism, which she developed a year later.[69]

Beginning in near silence, *Biogenesis* gradually progresses into a drone sound resembling a sonification of black holes,[70] a vibration that transitions into a long spatial sound that overlaps with the sounds of the heartbeats. The heartbeats are not immediately recognizable and at first appear to be deep electronic beats mixed with interference. The first set of heartbeats—that of the father—evolves slowly so that its metamorphosis into the heartbeat of the mother and the unborn child is almost imperceptible. "I manage to make one sound slide on the other," Radigue explained, "It's what gives the impression of almost flawless continuity."[71] The same happens in life, as recent biological research demonstrates, with the regeneration and continuity from cell to cell extending back billions of years. We often locate the beginning of a new life at the sound of a heartbeat and the completion of cell division, but we now know that all cells are formed by the fusion of other cells.

The baby's heartbeat is set at the end of the piece in a higher tone. This sound, which was recorded when the baby was almost due, competes with the complex sound world of the mother's interiority: her heartbeat, her digestive growling, and her breath. Life is part of the piece not only because the sounds are from the heart but also because Radigue considers sound itself to be "an autonomous life that needs to be respected."[72] This is why in her work, the flow of rhythms is slow and intuitively guided.

In his 1973 review in *The Village Voice*, after the performance of *Psi 847* (1973)—which, like *Biogenesis*, is dedicated to her daughter and shares its minimal materials and intimate slow-evolving style—the composer Tom Johnson reflects on Radigue's music:

> There is something very special about the music of Eliane Radigue, but after thinking about it for almost a week, I still can't put my finger on what it is. Is it the intimacy? . . . Is it the sheer efficiency with which it accomplishes so much with so little? . . . Is it that Radigue sustains her minimal material for minutes without ever repeating herself or becoming boring, and yet without ever leaving the restricted area within which she works?[73]

Simplicity and subtlety are the charm of Radigue's work. What is interesting about the use of heartbeat sounds in *Biogenesis* is her ability to manipulate the sound, with fade-ins and fade-outs, but without changing their essence. They remain what they are but also bring with them something different, a sensitivity to the passing of time and the continuation of life, the evolution of the cosmos to the development of life on Earth, all in a simple but continuous beat. In Radigue's music, as Emmanuel Holterbach has noted, "the body vibrates in a space that vibrates, and time folds in on itself to take on hues of infinity."[74] Radigue, who is in her nineties, no longer uses her APR synthesizer but continues creating compositions in collaboration with other musicians that contain beatings, breathings, flutterings, and vibrating tones. The main factors that have defined Radigue's music are space, time, and silence.

Space is present in *Biogenesis* not only in her use of reverberations and space resonance but also because the heartbeat sound she modulates is not easy to identify. Her sustained tones and low-frequency immersive landscapes create a sense of vastness and timelessness that transports us to outer space. Her deep resonating tones can mimic the rumblings and vibrations associated with celestial bodies and space travel. Her work, made of "quasi-cosmic vibrations," reminds us that life is a chain of cosmic pulses, a cosmic transmission, an intricate sonic phenomenon that originates from the processes of the universe.

3

ELECTRICAL

Although we associate electricity with lighting, power grids, power outlets, and electronic devices, electrical currents are produced by cells within our bodies, and they control everything, from our heartbeats to our perceptions of sight, smell, and sound, as well as movements and thoughts. Life is nothing if not electrical, an unseen force that surges and flows, spreading vitality. A coordinated sequence of electrical signals of oscillating waves is essential for the heart to function properly. The electrical impulses in our heartbeats start in the sinoatrial node, the pacemaker of the heart, which sets a normal rhythm of 60 to 100 beats per minute. They then travel through the atria and the upper chambers of the heart, prompting them to contract and pump blood into the ventricles, keeping this vital liquid ever moving. The entire heartbeat is a continuous loop of lub-dubs, of systoles and diastoles yearning for life.

Many artists have been inspired by the electrical activity of the heartbeat, as it incorporates rhythms, patterns, and repetitions. In this chapter, I concentrate on the work of Heinz Mack, Jean Dupuy, Mark Boyle and Joan Hills, and Diana Domingues, who have all used the heart's electrical signal to produce pieces that combine cardiac harnessing technologies, such as electrocardiograms, stethoscopes, and sensors, with light, vibrations, projections, and virtual reality (VR) to reveal multiple possibilities and meanings of the movements of the heart and its electrical energy.

Heinz Mack, in the late 1950s, began deploying cardiograms and incorporated them in light-based structures to reveal the vibrational energy of life. Later experiments merging art and science, like those of French artist Jean Dupuy, led to one of the first installations that made heartbeats visible through a combination of materials that included light and dust. At the same time, the first performances incorporating EKGs and heartbeat sounds were being pioneered by Mark Boyle and Joan Hills, who in their light shows used heartbeats and other bodily functions and fluids to expose the mechanisms of the body. Further developments in technologies later enabled experiences that immersed participants in their own hearts and bodies. Heartbeats, as electromagnetic waves detected by sensors, became more than guiding and biofeedback elements. In the hands of Brazilian artist Diana Domingues, they became part of an electro-trance ritualistic experience in which electricity had a conductive role in creating new conscious experiences.

VIBRATING LINES: HEINZ MACK

Entering a white cube, you are immediately attracted by a pair of large canvases depicting oscillating continuous lines marked in a single color in a single space. The curves remind you of something: the graphic tracings made by the stylet of an electrocardiograph. Working a century after Étienne-Jules Marey and Auguste Chauveau first published a series of cardiac graphs (1861–1863), the German artist Heinz Mack created *Cardiogram of the Cyclopes* (1961–1962) and *Cardiogram of My Heart* (1962; figure 3.1). These works are part of a series of captivating kinetic reliefs, paintings, and drawings composed of pulsating lines that Mack has made since the 1960s.

Born in the town of Lollar in 1931, in the Republic of Weimer, Mack moved in 1944 to Düsseldorf, where he graduated from the Kunstakademie Düsseldorf and was awarded a degree in philosophy from the University of Cologne, in West Germany, in 1956. The following year, he cofounded ZERO with Otto Piene and Günther Uecker, the renowned international post-avant-garde group, active from 1957 to 1966, whose zeitgeist was the return to ground zero for the transformation of arts after World War II.[1] ZERO's name was inspired by the moon landing and particularly the

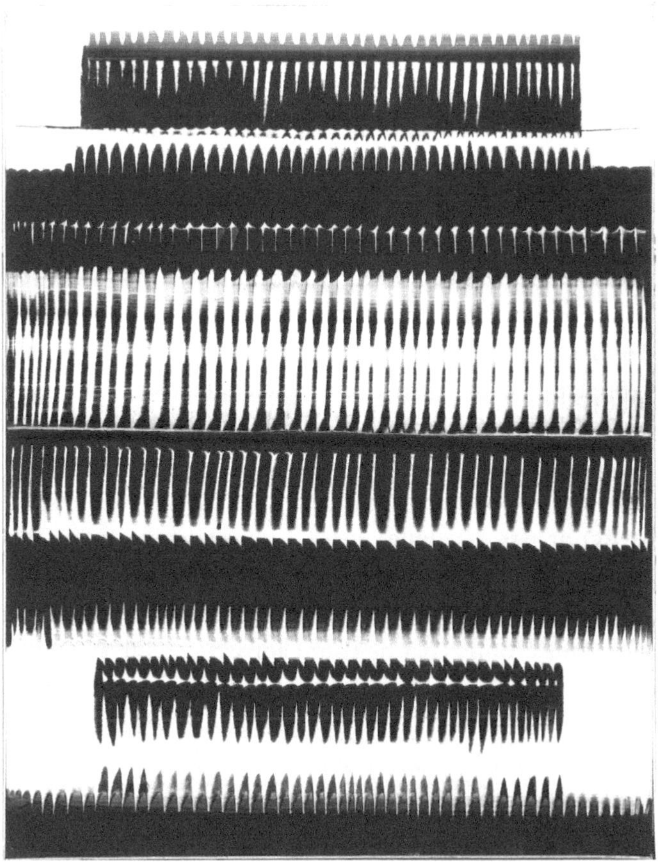

3.1

Heinz Mack, *Cardiogram of My Heart, No. 2*, 1962. Synthetic resin on nettle, 180 × 140 cm. Courtesy of Archiv Heinz Mack | VG Bild-Kunst, Bonn.

rocket's launch countdown: "10, nine, eight—was where we took the name from," Mack said.[2]

The group proposed alternatives to both realism and postwar subjective movements, such as art informel and abstract expressionism. Instead of the usual colorful and emotional flow drawn on by these movements, ZERO developed a monochrome and kinetic language focused on materiality. ZERO began with a humble magazine and a series of small exhibitions in their studios but went on to evolve into an international network of artists. As modern technologies and industrial materials were perceived as a male domain at the time, the group consisted mainly of male artists, such as Yves Klein, Lucio Fontana, and Jesús Rafael Soto, although female artists like Yayoi Kusama, Nanda Vigo, and Rotraut Klein-Moquay were also part of the group.

Interested in the relationships between nature, science, and technology, ZERO artists developed structures exploring light, space, and movement. In the case of Mack, these elements were interconnected parallel lines and grids. Through a simple but significant vibratory language, Heinz Mack found that he could construct a physical response to the highly gestural and emotional character of expressionist tendencies. Focused on the transformative power of light and the energy of vibration, the German artist broke binaries of abstraction and representation. He found in the monochrome a "pulsating and luminous space that makes the viewer sensitive to the energy of art and nature."[3]

Mack is today recognized as an international pioneer in op art, kinetic art, and land art. Throughout his career, he has continued to explore light, lines, and grids in his work while his sculptures, drawings, and paintings play with temporal superimpositions that demonstrate how his artistic production is not linear but rather akin to a circulatory system:

> In my body of work there is no chronological development. Early work, mid-career, late work; this linear thinking is a completely foreign concept to me. I subconsciously revisit certain things I've done before that I can develop further. This circular thinking gives the work a certain cohesion.[4]

Mack has dedicated his entire career to exploring the visual materialization of vibration. Vibrations are a universal rhythm that stretches and resounds from the farthest reaches of the universe to our bodies' organs, as well as to our own thoughts and feelings. Everything in the universe, including those elements that appear to be static, are in fact in motion, vibrating and oscillating at varying frequencies. Our bodies, which are composed of cells and atoms, are vibrating energy. The dance of our hearts, the lub-dub sounds that travel through our chests, the flow of watery matter within our bodies, and feelings of love and anguish within our hearts are all manifestations of vibration. The wiggly lines that register on a cardiogram are a visual representation of the heart's electronic and kinetic vibrations. Mack was fascinated by the linear movements and dazzling spectrum possibilities of the cardiogram. That's why he plays with movement, light, and space in his cardiograms or "dynamic structures" to bring vibration and structure to a "static" monochromatic surface. As the curator Edouard Derom has noted, "Mack qualifies a structure as 'dynamic,' implying thereby that it possesses not only movement but also energy, that is, the force that produces the phenomenon of motion."[5] In Mack's work, vibration, movement, repetition and rhythm are combined in a way that each stroke remains in its ontological place.

Cardiogram of the Cyclopes (1961–1962) and *Cardiogram of My Heart* (1962) are early paintings composed of monochromatic, straight but discontinuous lines of variable dimensions. Painted with synthetic resin in various tones of gray, they are all arranged vertically and horizontally, following Mack's structure-based concept of painting.[6] The interplay of lines positioned in repetitive sequences and painted in tonal gradients creates a rhythmic pulse or vibration, an optical energy that recalls a cardiac cycle.

> The pattern of such lines is infinitely variable; the principle involved is the simultaneous and recurring sequence of the lines. (These lines need not be exact.) Every zone in between two lines displays an immensely rich color—form structure; accidental creation is given a real chance—the lines arrange themselves. Individual parallel zones gradually transform themselves from zone to zone, while at the same time they retain their distinct but mutual character—in this way they are brought into vibration.[7]

An outstanding relief work, which plays with light and movement to create heart-like rhythms, is *Little Diagram of My Heart* (1967). This handmade piece presents several vertical indentations embossed in aluminum. As the metal is reflective, it produces a rhythmic vibration upon contact with daylight or artificial light. Equally effective is *Heart Rhythm* (1971; figure 3.2), a square metal foil work in which a dynamic jagged line is produced by the reflection of white light on its surface. This induced refraction flickers in space, creating an apparent motion. Mack defined this effect as "optical energy," which reproduces the reverberations of a heart tracing. These refractions move according to the position of the viewer and create a form of interaction between the observer, the work, and its surroundings.

Mack's cardiogram works, including his paintings and reliefs, are visible reminders of two forms of energy essential to life: heartbeats and light. Mack's fascination with both lies in the fact that they are each a physical energy wave and a metaphysical phenomenon. Both create a sort of static energy that Mack calls the "resting restlessness": an oxymoronic aesthetic form "experienced by our eye" that, Mack says, is "scarcely perceptible, a contrast to the rhythm of the heart; it is movement that destroys itself."[8] It is an anomaly produced by repetitive patterns that suggest heartbeats as a self-contained energy. In these works, Mack shares the intention of creating "motion without motion" or "static movement" with other avant-garde artists, such as Giacomo Balla and Marcel Duchamp mentioned in chapter 1.[9] Mack applies light, lines, and color as a means to create effects that interrogate not only the materiality of objects but also the dialectics between art and technology.

His less well-known works *Rhythm of Seeing* (figure 3.3) and *Cardiogram for Africa*, both from the *Coronar Edition* (1990) series, depict variations of cardiograms alongside electrocardiograms of his own heart. In these pieces, two different forms of energy collide: one symbolic but mechanical, driven by the movements of the hand, and the other corporeal and electrical, produced by the motion of the heart. Mack focuses on rhythm—namely, the strong, regular, and repeated pattern of movement that produces a mechanical effect—but also refers to the organic rhythms found in the body.

Rhythm of Seeing (1990) is a palimpsest-like serigraph. The background of the piece is a grid and a dotted line pattern, over which the artist has superimposed sequences of lines bounded by two electrocardiograms of his own heart. Mack calls these drawings "graphical energy fields," which

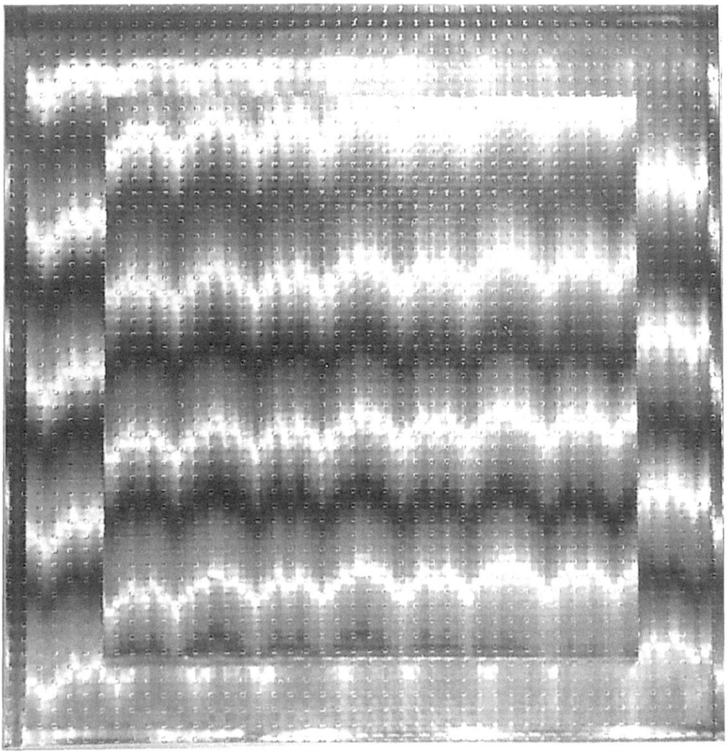

3.2

Heinz Mack, *Heart Rhythm*, 1971. Acrylic glass, metal foil, wood, stainless steel, 50 × 50 × 6 cm. Courtesy of Archiv Heinz Mack | VG Bild-Kunst, Bonn.

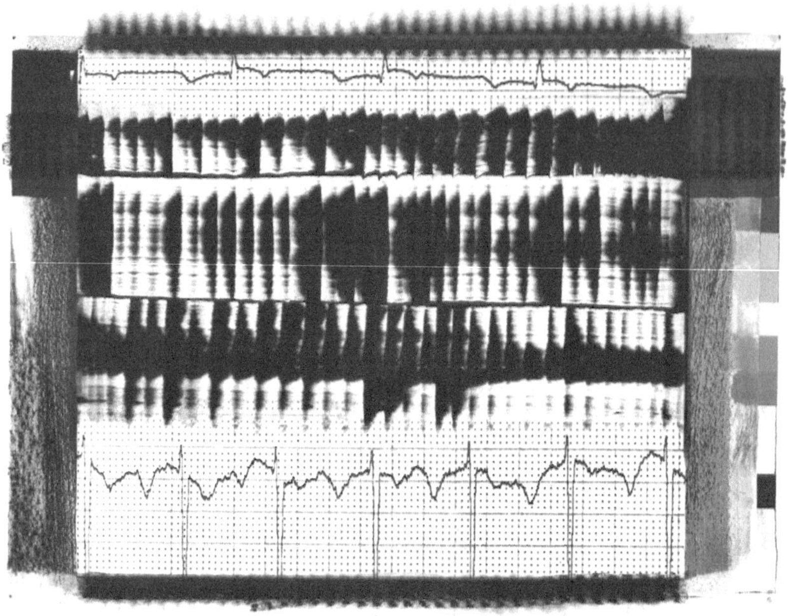

create vibration through repetition and difference, shadow and light. In this experimental work, vibration is represented as a life force that flows through and regulates the body, an oscillation that rises and falls, that repeats but also changes, as is the case with all rhythmic movements.

The artistic method in this case is an open mechanical arrangement. The mechanics of artistic activity gets stimulated and directed by the sensitivity of the artist's hands—hence it escapes the danger of mechanical repetition, which continues to be uncreative. In principle, all parallel zones have the same intensity of presence.[10]

3.3

Heinz Mack, *Rhythm of Seeing (Coronar Edition)*, 1990. Sixteen-color screen print on copperplate, 105 × 136 cm, laid paper. Courtesy of Archiv Heinz Mack.

Mack once declared that he was identical with what he makes, and *Rhythm of Seeing* is a piece that draws together different elements of his production: the early optical grid graphite drawings that recall EKG grid paper, the monochromatic cardiac-looking lines, and his own EKG tracing, creating a sort of rhythmic depiction of his own life and practice.

Mack's idea of repetition in revisiting past forms in his work and his use of vibrating lines aligns with Gaston Bachelard's idea that the self has an "undulatory" nature. This is because life is not linear but rhythmical, created by mentally organized, continuous, and discontinuous instants.[11] Bachelard further noted that "if a particle ceased to vibrate, it would cease to be. It is now impossible to conceive the existence of an element of matter without adding to that element a specific frequency. We can therefore say that vibratory energy is the energy of existence."[12] Extending on the topic, Henri Lefebvre affirms, "Everywhere where there is an interaction between a place, a time and an expenditure of energy, there is rhythm." Rhythmicity is a universal phenomenon found in heartbeats, waves, molecules, and life cycles. That is why Mack's works play with rhythms, repetitions, movements, and interferences of linear and cyclical processes that muse on vibration and light.[13]

In Mack's work, we find a cyclical and circadian repetition guided by the effects of light and the cycles of the artist's own life incorporated in a linear sequence or imposed structure, in which the biological and the mechanical, the intellectual and the physiological, the ontological and the physical are interrelated. As Bachelard noted, the "physical or physiological phenomena would always teach us to submit to time and be an object among objects."[14] And it is in this dualism between subject and object that Mack's *Cardiogram* works are situated; cardiograms are neither a subject nor an object but vibrational energy.

This same rationale is found in *Cardiogram for Africa* (1990), in which his own cardiogram is accompanied by dynamic line structures in a desert-like brown palette. The choice of colors recalls the shimmering heat on the horizon in the desert that is commonly described as vibrational. Mack has a profound connection to the Sahara, where he has spent time since the 1950s studying the effects of light in the vastness and on him and his works. In the desert, Mack usually wears a custom-made silver suit in order to blur the boundaries between self and material, between work and life, like his predecessor Étienne-Jules Marey, who was also interested in movement, space, and time, and considered the graph to be an objective "language of phenomena themselves." Mack recognized the graph as an experimental medium and has referred to his graphic drawings as "the language of my hands" and "the grammar of my art." He believes that they connect with an energy field that flows according to the current of his ideas and feelings.[15] Thus, these art pieces or "graphic energy fields" represent not only vibration but a vital sign and an artifact: it is electrical and mechanical, subjectivity and objectivity, and an inner relationship between the inside and the outside, reality and illusion, philosophy and physics.

It is within this convergence of the visible and the unseen that a unified reality emerges. Abstraction is no longer essential for creating a meaningful collision; now the ontology of being, if any, is physically embodied with motion and direction in an immobile vitality, in which repetition finds its difference and life given materiality in a cardiograph. In Mack's cardiograms, the heartbeat is a dialectical mark of light and contrasts, peaks and valleys, a line that oscillates between subject and object, where the material and the immaterial, and the artificial and the organic meet in a constant creative movement. It is a vibratory energy, the energy of existence.

THE VISUAL ENERGY OF HEARTBEATS: JEAN DUPUY

Inside a black box, you see red dust dancing energetically, following the rhythm of your heartbeat: it floats in the space it inhabits but quickly becomes inert until the next heart forces it back to action. Working a century after Laennec's achievement of making hearable the lub-dub sounds of the heart, Dupuy's *Heart Beats Dust* (1968) aims to make visible the heartbeat using a stethoscope. The French artist always claimed that he used technology only to show things that were invisible.[16] *Heart Beats Dust* is one of his best-known works and among the earliest pieces to merge heartbeats and light in a three-dimensional work. The sculpture consists of a dark rectangular plywood box with a 60-centimeter glass window at eye level. Inside the box, an infrared light located in the roof illuminates a pile of Lithol Rubine—a bright blood-red powdered dye of low density that can remain suspended in the air for long periods.

While Heinz Mack was interested in revealing light and vibration in his two-dimensional structures, Dupuy was more interested in making an interactive dancing sculpture using light and vibration to reveal the cardiac rhythms of participants. The first prototype of his *Heart Beats Dust* installation was activated by taped heartbeat recordings to make heartbeats visible, but Dupuy was unsatisfied with the result. So, for his second version, Dupuy used a stethoscope to retrieve the visitor's cardiac rhythm and played it back to a coaxial speaker mounted directly below a rubber membrane on which the red powder rested. Sound waves from the heart activated the dust so that once it came in contact with the infrared light, it created a pulsing red, cone-shaped sculpture. The work was later retitled *Cone Pyramid (Heart Beats Dust)* (figure 3.4) to acknowledge this form. Dupuy illuminated the hidden secret of the continuity of our existence: the heartbeat, that recurrent movement whose lub-dub sounds, produced by the rhythmic closing of the heart valves, create a celebratory dance of life. Thus, *Heart Beats Dust* turns a simple wooden box into a space where lights and heartbeats (two electrical waves) converge and where an intimate connection and fusion is made between the mind and heart, and the biological and the technological.

Born in Moulins, France, in 1925, Jean Dupuy began his career as a figurative painter in Paris. Dissatisfied with his own pictorial practice, however, he decided to destroy all his canvases by throwing them into the Seine, and, in 1967, he moved to New York in search of new ideas. Like his fellow national Éliane Radigue, Dupuy found the United States to be an exciting and experimental environment. There, a number of artists, through the initiative Experiments in Art and Technology (E.A.T.), had begun collaborating with engineers to explore the synergy of arts and technology in cybernetic and other movement-based works.[17] Inspired by New York's technological focus, Dupuy started an ambitious series of works that mingled the relations of space, art, and life; nature and technology; and sound and sight.[18] With the aid of microscopes, telescopes, periscopes, and stethoscopes, he strived to show phenomena that were invisible or imperceptible to the senses. This all started from his contact with E.A.T.

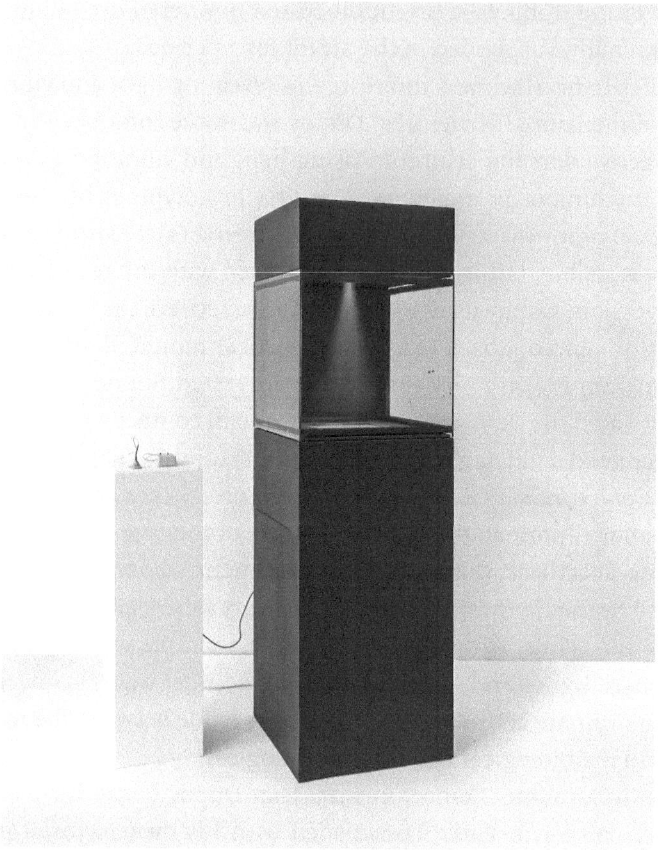

3.4

Jean Dupuy, *Cone Pyramid (Heart Beats Dust)*, 1968–1969. Stethoscope, spotlight, amplifier, wood, glass, red pigment, 190.5 × 56 × 57 cm. © ADAGP, Paris. Photo by Fabrice Gousset. Courtesy Loevenbruck, Paris.

E.A.T. evolved from Billy Klüver and Bob Rauschenberg's *9 Evenings* (1966), which was presented at the Armory in New York with an ensemble of artists, dancers, filmmakers, composers, and engineers revealing motor aspects of the human body.[19] I was in New York for four months in 1966, but I missed the event. However, poet Claudio Badal introduced me to the E.A.T. directors: engineer Billy Klüver, and artist Robert Rauschenberg. . . . [In 1968] E.A.T. proposed to make an exhibition of works envisioned by an artist and executed by an engineer. The Brooklyn Museum hosted the exhibition *Some More Beginnings* (1968) as well as a competition. Six hand-picked engineers appraised the works. I won the competition with my *Cone Pyramid*. Concurrently, Pontus Hultén, then-director of Stockholm's Moderna Museet, organized *The Machine as Seen at the End of the Mechanical Age* (1968) at MOMA. After winning the Brooklyn competition, I had to redo a second *Cone Pyramid* because the winner was immediately admitted to the more traditional show at MOMA: Glory![20]

Cone Pyramid (figure 3.5) is the experimental sculptural dancing machine that positioned the French artist at "the start of participatory tech-art exhibitions in the United States and France."[21] Hultén, who was one of the few arts people on the jury, which was otherwise comprised entirely of male scientists and engineers, declared Dupuy's interactive piece to be "a new form of cooperation with nature. A sensitive collaboration between natural forces within and outside the human body."[22]

Dupuy and his collaborators had succeeded in amplifying the heartbeat and making it visible by simple means, something that many artists and engineers, including Klüver himself, had previously attempted. A few years earlier, for example, Yvonne Rainer had consulted Klüver, who was the intermediary between artists and technicians at Bell Labs in New Jersey. She wanted to use her heart rate for *At My Body's House* (1964), but the engineer told her that this was not feasible, so she ended up using a contact microphone instead to amplify her breathing.[23] *Heart Beats Dust* was a simple but innovative work for the time and rich in meaning. More than making a link between human and nature, organism and machine, Dupuy's work was a flux of physical and physiological energies, which referred to our cyclical ephemeral nature through the combination of three deeply symbolic materials: heartbeats, light, and dust.

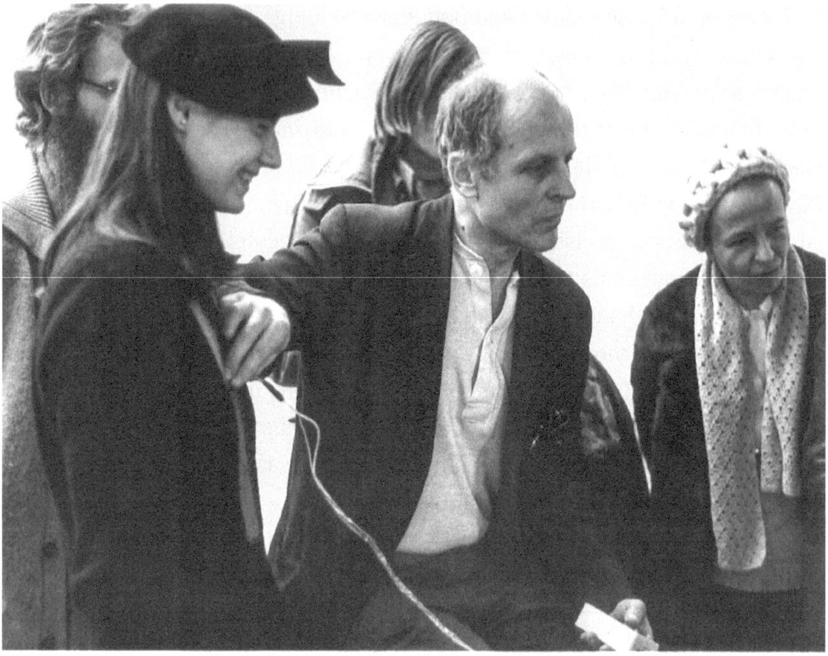

For millennia, light has been a metaphor for life, a symbol of knowledge, and is now known to be the basis of a profound relation to heartbeats: they are both electrical waves that pulsate. From the earliest times, light is often referred to in religious texts as the beginning and source of life: "Let there be light." Heartbeats and lights have been historically interconnected as signs of vitality; if there is a throb in the heart, there is life and, therefore, "brightness" in someone's eyes. The light that enters the retina is converted into electrical signals by the light-sensing cells, which turn off when life ceases. And then there is dust. These particles

3.5

Jean Dupuy activates *Cone Pyramid (Heart Beats Dust)*, 1968. Presented at the exhibition "Für Augen und Ohren," Akademie der Künste, Berlin, 1980. © ADAGP, Paris. Courtesy Loevenbruck, Paris.

cover most of the planet and can be found in space. Like light, it refers to the cyclical nature of life: to the origins and to the ending of vitality, as well as to everyday life and to eternity. This idea can be found in the words of the Bible, "from dust to dust" and "dust thou art," which refer to both being and nothingness. Furthermore, as all the elements on Earth and therefore our bodies were generated in stars, we are stardust.[24] So, dust is a trope for the beginning and the end of life, and as in *Heart Beats Dust*, it creates bodies, but it also replaces bodies. Like dust, the heartbeat is a sign of animation and nonanimation, first heard from a fetus in the mother's womb and searched for to confirm the end of a life. Heartbeats, dust, and light mark the passage of time and cross the boundaries between the living and the dead. Thus, the three main materials selected by Dupuy represent life and operate poetically.

In *Heart Beats Dust*, Dupuy used a beam of light to illuminate the invisible movements of the vital organ inside a box containing a fine red powder. Owing to this color, these fine particles can be associated with the heart and their movements with the force of the living body. Thus, the work illuminates an inner vital dance, a show that lasts twenty-four hours and that does not stop for years until life meets death and becomes dust.

Owing to its significant temporal and spatial connotations, dust is a material that many artists have employed, including Marcel Duchamp, Yves Klein, and Robert Rauschenberg. Duchamp, from whom Dupuy's sensibility stems, was one of the first artists to use dust as an artistic material. Fascinated by a year's accumulation of dust on *The Large Glass* (1915–1923), he decided to permanently affix the fine powder on an area of his work's translucent surface. Before being cleaned away, the entire pile of dust was captured in Man Ray's photograph *Dust Breeding* (1920).

Like Duchamp, Dupuy has stated that his idea of using a pulsating powder derives from various experiences that made him aware of the physical and cumulative properties of particles:

> I lived in Bowery, south Manhattan, where the tramps lived; It was then that I happened to meet the manager of the American company Celanese Co., who gave me three cubic meters (180 plates, 200 × 90 × 0.6 cm) of polyethylene plates in six colors. . . . Once the piles of plastic were at my studio, I became a real housekeeper, because the electrostatic charge from polyethylene attracted all the dust

from the workshop. I spent a lot of time wiping the dust off. This action made me realize that dust had its weight. One day during a movie at a cinema, I also noticed that the dust in the room allows you to see the rays of light that project the images on the screen. I then imagined making a sculpture of dust. So, I needed a light projector and a very low-density pigment. Finding the motion to stir the pigment took the longest.[25]

As Michael Marder has noted in his book *Dust*, the act of dusting is a self-defeating occupation because it deals with a material that, by its nature, is unsettled, spreading through the air and gravitating toward other objects.[26] Dupuy's use of a dust-like, red-colored powder is associated with an articulation of space and blood that turns an invisible throbbing into a three-dimensional form. The "dancing cone," that image that mesmerized many of his contemporaries like James Turrell and Roland Barthes, is what inspired and guided Dupuy to select a light beam to illuminate heartbeats.[27] Seen in a ray of light, "floating in the air, dust draws our attention to space, to the in-between we habitually overlook."[28] Dupuy used dust's capacity to illuminate and reveal things that are not visible and disregarded, such as heartbeats.

The lub-dub sound of heartbeats activates the sculpture, causing dust to erupt. As Dupuy noted, this sound became the main "driver" of the work, providing the "heart a second function: that of showing an observer a direct, vibrant, and living reflection of himself."[29] The person who activates the work becomes a participant, although mostly as an observer. But in this observation of the self, there are no traits or genders, only dust that beats, and the sound of heartbeats transmitted by the stethoscope (Laennec's medical device that has shaped our acoustic perception of the body) enables the experience of exploring our interiority in an external space.

Dupuy's interest in the energetic and sonic character of the heart can be traced back to childhood visits to his grandfather who was a doctor. He had many medical instruments in his storage cabinets, including a Laennec stethoscope with which the young Dupuy heard for the first time the sounds of his own pulsations.[30] Dupuy's pulse-based dancing sculptures can therefore be related to the musical and dance forms used by Laennec, Marquet, and other physiologists to refer to the pulse (described in chapter 1). Pulse and light, like dance, are energies that draw lines in space and time. Dupuy's heartbeat pieces are guided by rhythmic waves of life that make visible variations of energy in the flow of movement.

During the late 1960s and '70s, Dupuy combined different media and art forms, such as dance, choreography, and cinema in his installations and also in other events. He organized several evenings in his own apartment as well as in iconic locations such as The Kitchen, the Whitney Museum, and the Judson. Before *Heart Beat Dust*, he created a dance work in which he materialized his childhood memories of hearing the heartbeat:

> [After] seeing the rays that cross the cinema hall to project the images on the screen . . . I would then collaborate with a young woman in a dance show that she was preparing at Columbia University, at the theatre of the Casa de Italiana. I had proposed to place, on the stage of the small theatre, some stairs seen in profile in relation to the room, for the simple reason that we rarely have the opportunity to see a person going up or down a stair from a side view. . . . So, she had to go up the stairs, disappear and go back down on stage. Speaking then of rhythms to accompany the action, I came up with the idea (by chance?) of using heart rhythms. There was a 33-rpm record of heart murmurs (at Sam Goody's, a record store in New York). These are the circumstances, the sequence of circumstances, that gradually led me to this sculpture.[31]

Another choreographic work, *The Visual Energy of Sound* (1968), was an 80-minute dance concert Dupuy staged at the Judson Church Theater.[32] Dupuy's event was divided into two phases: "La Course" juxtaposed two elements in movement, body, and music, in a space structured by stairs.[33] "The Visual Energy of Sounds" consisted of recordings of the noises of several objects (shells, ladders, brushes, radios, irons, hair dryers, fans, etc.) and the gestures of people using those objects. It was a work in which the body, musical rhythms, and movements were the protagonists and the exploration of sound as an energy produced concrete as well as abstract effects.

Following on from *Heart Beats Dust*, *Chorus for Six Hearts* (1969; figure 3.6) was another performative work conceived for *The Machine and Nature*, a special public program that was part of Hultén's exhibition at the Museum of Modern Art. It was displayed in the museum's Founders Room on January 15 and 16, 1969. *Chorus for Six Hearts* is one of the first pieces

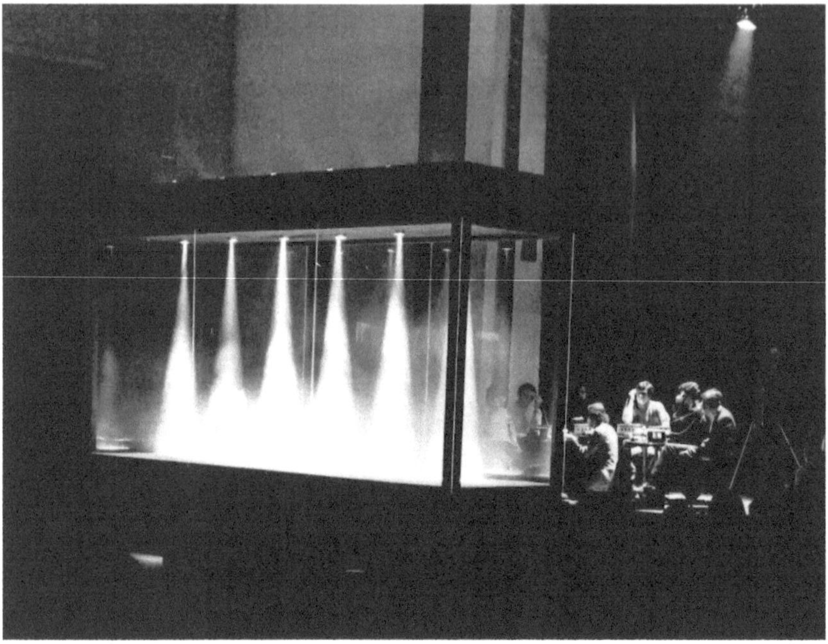

that made visible a collective of heartbeats. In a dark room, in front of six copies of the same system used for *Heart Beats Dust* (1968), six performers pressed contact microphones against their bodies—chests, cheeks, throats, necks, and mouths—broadcasting not only the visual energy of their heartbeats but also of many other bodily sounds. These vibrations created an arrhythmic dance of flickering lights and dust that was guided by both the phenomena and by Dupuy as he manipulated the volume of the signals coming from each performer. The *New York Times* described Dupuy's 45-minute-long audio-visual action as "very soothing, very quiet,

3.6

Jean Dupuy, *Chorus for Six Hearts*, 1969. Festival Sigma, Théâtre de l'Alhambra, Bordeaux, France, 1969 © ADAGP, Paris. Courtesy Loevenbruck, Paris.

mildly absorbing, mildly mesmerizing, and the ending, in which the performers laugh wildly (but still softly) into their microphones, was a relief."[34] Other participants were fascinated by the fact that such a primordial sound, reminiscent of ancient drum beats, was a physical phenomenon perceived by means of technology.[35]

Dupuy's observations of the heartbeat culminated in *Pulse* (1970). Presented at the Whitney Museum's *Motion Pictures*, the 13-minute silent Super 8 film presents dancing cones produced by the pulses of three different people taken from the wrist, jugular vein, and crook of the elbow. Heartbeats were shown to be an invisible energy guiding a dance by setting a rhythm of interaction in which the participant becomes a producer, performer, accomplice, and observer.

Dupuy, who died in Nice in 2021, was a pioneer of work that use art and technology to reveal the hitherto invisible. At the core of all these pieces that Dupuy created with heartbeats is the use of technological elements that combine the machine with the body to create a conscious experience of life. The works are the result of a constant but not homogeneous temporal wave that moves through space, always accompanied by a transfer of energy. This wave that emanates from the body and is turned into motions, light, and sound reminds us that inside our skin, there is something pulsing, dancing, following the pace of our life.

LIGHTING BODILY ENERGIES: MARK BOYLE, JOAN HILLS

Imagine that you are in a dark room where you can hear a constant heartbeat experienced both in sound and visual forms. The constant rhythmic ticking is accompanied by a large number of materials and energies, such as those of our regular bodily functions. This is the main thread of *Son et Lumière: Bodily Fluids and Functions*, a performance by Scottish artists Mark Boyle and Joan Hills, presented for the first time at Liverpool's Bluecoats Arts Centre in 1966. Similar to Dupuy, Boyle and Hills created a performance that reveals the mechanisms of the body but in a directly transgressive way. The light show involved the production of bodily secretions (phlegm, nasal mucus, urine, sweat, blood, earwax, tears, sperm, gastric juices, saliva, and vomit), with the visual examination of substances stained with dye using projected microscopic images and the amplification and visualization of heartbeats and other bodily sounds. In the countercultural spirit of the times, with its increased corporeal and sexual candor, anatomical materials were explored, and the limits of the body were contested. In his text, "Atlas and Manual," Mark Boyle described the eleven instances of the performance:

FLUID ACTION | SOUND REACTION | PROJECTED ONTO A SCREEN

1. Catarrh Coughing up | Contact mike on throat | Examine with microscope dyes.
2. Snot Blow nose, etc. | Contact mike on nose | Examine with microscope dyes.
3. Saliva Kiss to arouse saliva possibly eat | Contact mike in mouths of kissers or on throat if eating | Examine with microscope dyes separate starch, etc.
4. Earwax | Extract | Melt and examine.
5. Tears | Extract Tape—children crying, adults howling | Crystallise tears in microprojector.
6. Urine Piss in can | Contact mike on can | Sugar test on urine.
7. Sweat | Exercise furiously | Breathing and heartbeat | Examine with microscope dyes.
8. Blood | Extract with hypodermic. Loud breathing and heartbeats | Examine with microprojector.
9. Sperm | Extract by copulation or masturbation | Attach electrocardiogram [*sic*] and electroencephalogram [*sic*] | Televise oscilloscopes of equipment, project using TV projector. Maybe silence, maybe climactic heart and breathing sounds | Project live sperm in microprojector.
10. Gastric Juices. Swallow sponge on the end of a string. Pull up | Contact mike on stomach | Examine with microprojector.
11. Vomit Take emetic | Contact mike on stomach and throat | Examine with microprojector.

NOTES

12. I think moral shock tends to be a barrier to perception. It is therefore perhaps advisable to leave out any sections that might cause moral shock to the particular audience.
13. The entire piece may be performed with the continuous rhythm of breathing and heartbeats.[36]

Lights and the constant sound of heartbeats were the main driving forces in *Son et Lumière: Bodily Fluids and Functions*. During the premiere of the performance, only three stages were accompanied by heart and breath sounds: when the performers exercised furiously to produce sweat, when blood was drawn from their bodies using hypodermic needles, and during the phase in which sperm was extracted by copulation or masturbation. After seeing the powerful effect of the heart's drumlike reverberations, subsequent presentations in London were adjusted slightly and accompanied by the constant sound of heartbeats. These sounds were soft or loud depending on the phase, emphasizing the relationship between physical activities and cardiovascular responses.

Using various technologies, *Bodily Fluids and Functions* reveals the functional aspects of the body in a liquid light show, with several gender implications at play. The work of this artist duo was shown only under Mark Boyle's name for many years because this option was seen as a more likely path to attract both notoriety and gallery commissions in a male-centered art world. Later, as light show performers, they became known as the Sensual Laboratory, and in the 1980s, they began working as the Boyle Family with their offspring. This latter designation was intended to stress the significance of each family member's contribution, as well as demystify the historical idea of the lone male artist genius and advocate for cooperation rather than individualism.[37]

Bodily Fluids and Functions renders a divide between feminine and masculine conceptions of the pumping organ. The heart has always been gendered: it is feminine because of its emotional significance and masculine because biomedicine has viewed it as a pump. In Boyle and Hills's *Bodily Fluids and Functions*, the feminine fluid and emotional image of the heart (the one that cries, has passions, and expresses emotions) is overshadowed by the masculine vision of the heart as a machine. The scientism and objectivity that obscured the heart's passionate feminine essence turned *Bodily Fluids and Functions* into a clinical performance.

Like nineteenth-century demonstrations of bodily functions using light to show the mechanisms of the body (described in chapter 1), Boyle and Hills in *Bodily Fluids and Functions* projected the body as an energy system to the audience. By revealing a grouping of corporeal (electric, kinetic, libidinal) energies and rhythms that examined performers from head to toe and from heartbeats to blood, Boyle and Hills presented an extreme portrait of nature and the mechanisms of the whole body. The body was transformed into a performance instrument illuminating fluids, rhythms, sensations, and bodily experiences.

The significance of the heartbeat fluctuated depending on the performers' actions, but it always demonstrated the interconnectedness of physiological processes. In the blood sequence, for instance, the lub-dub sounds were a reminder that the heartbeat is a muscular movement that keeps the vital fluid flowing throughout the body. In the sweating section, their heart rates were a signifier of physical exertion. And the semen sequence exposed the increased heart rate produced by libidinal energy when two strangers wired up to an electrocardiograph and electroencephalograph have sex on the floor, with their waveforms projected on a large screen behind them (figure 3.7). The cardiac rhythm was used as a symbol of vivacity and the body as a responsive machine.

Just as heartbeats were among a multitude of materials and energies in *Son et Lumière Bodily Fluids and Functions*, the performance itself was part of a series that also included different forms of liveness. *Son et Lumière for Earth, Air, Fire and Water* (1966) consisted of projections and sounds of chemical and physical reactions, and *Son et Lumière for Insects, Reptiles, and Water Creatures* (1966) was based on the inspection of animals and their sounds. Boyle and Hills's work sought to develop "new ways of seeing and presenting reality."[38] And as curator Andrew Wilson noted, by including in *Son et Lumière* the four elements that make up the universe, the living beings that inhabit it, the performances produced "a picture of the world and everything in it . . . a language of curiosity and metonymy, in which the part stands for the whole."[39] While this is accurate, in fact this work goes further. By showing that the components of life include both the biological (heartbeats and other physiological processes) and the symbolic, the artists created a consciousness not only of the world but also of the fact of living and its energetic processes.

The liquid lightshows pioneered by the Sensual Laboratory, which created a sort of *Gesamtkunstwerk* by combining different media and materials, became an important part of the psychedelic counterculture of the late 1960s and '70s in the United States and the United Kingdom. Artists

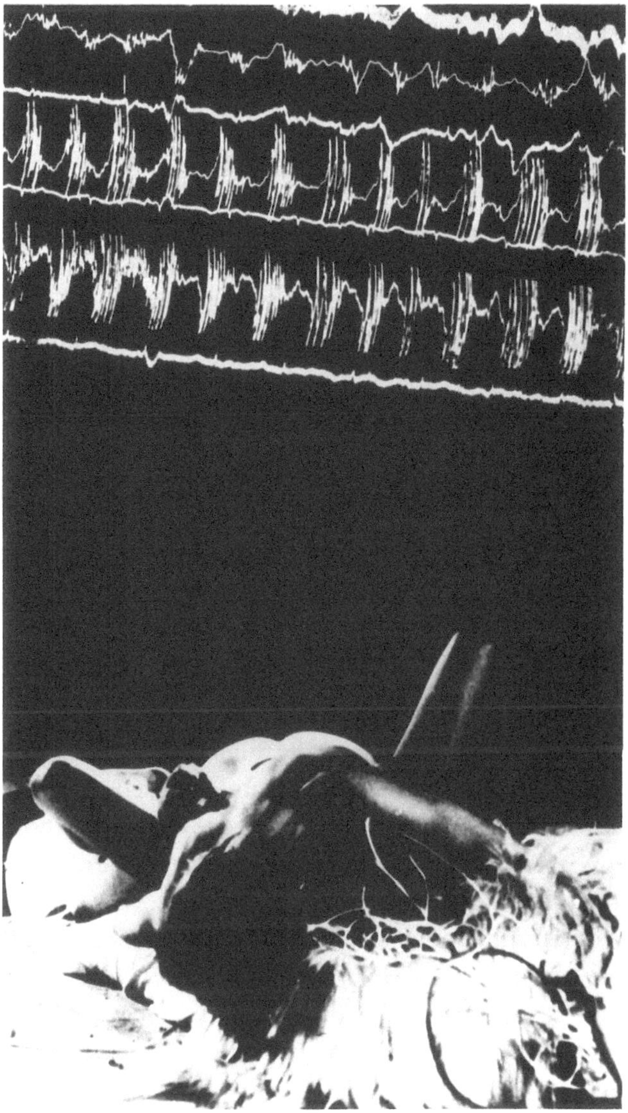

3.7

Mark Boyle and Joan Hills, *Son et Lumière: Bodily Fluids and Functions*, 1966. Performed at Roundhouse, London, 1967. Courtesy of the Boyle Family.

used a range of technologies, such as overhead, slide, and film projectors; modified slides; color wheels; films; and flashlights to intensify emotions and sensorial experiences.[40] The images in the light shows were mostly abstract forms in constant flux, swirling and changing color in time with the music. Boyle and Hills's pulsing images produced by heartbeats are a good example.

Light events like *Son et Lumière* were mostly shown in the United Kingdom in galleries and at festivals and underground clubs, such as the legendary rock and artistic hub UFO.[41] Figures like Lymbic System, Pink Floyd's lighting artist Peter Jenner, Peter Wynne-Wilson, and Mike Leonard were all active in the underground scene. While cardiac sounds and reverberations were incorporated by bands like Pink Floyd in some of their songs, such as "Heart Beat, Pig Meat" and "Eclipse," Boyle and Hills harnessed heartbeats live, and their large-scale projections of chemicals and bodily fluids were among the most experimental of the time.

During a tour in California, Boyle and Hills became aware of the main differences that distinguished their light environments from those of American artists, like the Joshua Light Show or Andy Warhol's Exploding Plastic Inevitable. While the Americans showed an overt virtuosity through their control and manipulation of materials for composing a show, the Scottish couple let the bodily fluids and substances flow naturally. They never forced the sequence or guided the course of the images, but the patterns always seemed synchronized with the pounding beat of the performers' hearts and the experimental rock music of their longtime collaborators, the Soft Machine.[42]

John Hopkins, the comanager of UFO, noted a bridging of the senses in the *Son et Lumière* events. He commented that Boyle and Hills's "combination of different sense inputs, imitated in some way the synesthesia you get with LSD."[43] One of these effects is an elevated heart rate. It is not surprising, then, that LSD-fueled visions created organic, hypnotic, and pulsating patterns similar to the projections presented in the shows that were also guided by cardiac rhythms.

Boyle and Hills's intention to create a theater that was not a representation of life but life itself finds its precedent in Antonin Artaud. As curator Andrew Wilson has noted, Artaud was an influential figure in the development of the UK counterculture and is "a vital chord" for examining the works of Boyle and Hills, and many other artists of their generation.[44] As Martin Esslin argues, Artaud was also "a believer in the healing power of the life force, the power of man's natural instinct as against dry-as-dust rationalism . . . he supported the heart against the head, the body and its emotions against the rarefied abstractions of the mind."[45] He believed that civilization turned humans into repressed creatures and that theater was a medium in which humans could liberate their instinctual energy. Like Artaud, by using high and low materials and a range of cultural products, Boyle and Hills tested hierarchical relations, as well as the static relations between artists and audience. Inviting participation to inspire new forms of interaction, they gave free rein to an unconstrained exploration of emotions and bodily sensations.

Boyle and Hills's intention of using the body and its fluids for confronting social conventions and for creating cinematic performances was mostly influenced by the "kinetic theater" and politically erotic work of American artist Carolee Schneemann. Boyle and Hills had their first direct contact with Schneemann's work when Mark Boyle collaborated as assistant producer of *Meat Joy* at the Dennison Hall in London in 1964.[46] Schneemann's colorful orgy involved naked bodies crawling on the floor, playing with paint, raw fish, and sausages, creating a wild feast of mixed bodies and materials. *Meat Joy* was inspired by Wilhelm Reich's sexual liberation theories, in which he conceived that individual and social neurosis was a product of unreleased libidinal energy. Another of Schneemann's erotic works was the film *Fuses* (1964–1967), which captured in image and sound her sexual activities with the composer James Tenney, creating a vibrant montage of images, colors, and abstract shapes. Similarly, the Boyle and Hills performances feature colors, shapes, and patterns that pulsate with the same erotic tone found in Schneemann's work.

In *Son et Lumière: Bodily Fluids and Function*, heartbeats, sound, lights, and corporal secretions combine with a burst of kinetic energy (figure 3.8). As the performers were in front of the projections, moving and generating plays of light and shadow, they also interacted with the images of waveforms and bodily fluids and became one with them. As Maurice Merleau-Ponty has noted, there is no other way of knowing the body than that of living it—namely, of feeling it, of performing it.[47] By projecting their own heartbeats, functions, secretions, and other physiological phenomena

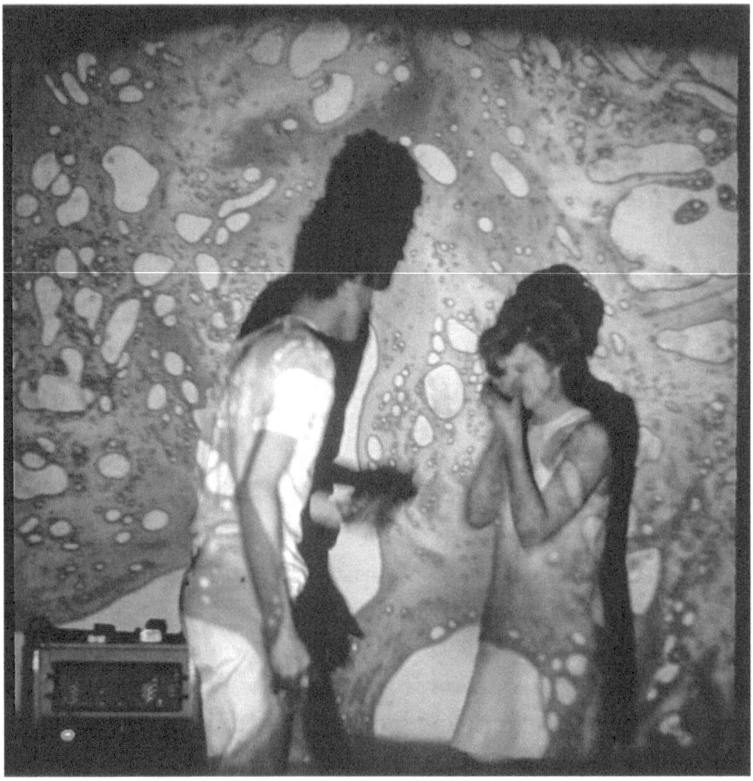

3.8

Mark Boyle and Joan Hills, *Son et Lumière: Bodily Fluids and Functions*, 1966. Performed at Bluecoats Society of Arts, Liverpool, January 1967. Courtesy of the Boyle Family.

out into the world to be seen, heard, and appreciated, Boyle and Hills developed a way to unify the two experiential modes of the body—the inside and the outside—by crossing their boundaries. The show, which displayed the body in its multiple forms in action, revealed the body as a whole, with its private and public aspects, including the heartbeats that make the body alive.

The heartbeat sounds in the performance were usually retrieved directly from the bodies of the performers, but on one occasion, in 1967, at the Jeanetta Cochrane Theater in London, the artists resorted to the use of an LP. At the end of the performance, one person in the audience approached Boyle and asked, "I don't wish to alarm you but are you aware that you have every known form of heart disease?"[48] The recording the artists had used was probably London's *Auscultation of the Heart* (1966), which had been initially marketed to medical students.

By combining the heartbeat and other bodily substances in *Bodily Fluids and Functions*, the light environment turned the body into a performing instrument of movements and sounds that drew viewers and participants to their own sensations and bodily experiences. The performance occupied an intermediate space between emotional experience and scientific exploration and observation of physiological activities. On the one hand, the body was perceived as a set of physiological and mechanistic variables (raised heartbeats, altered breath rates, and cold sweats), while on the other, the visual fluxing images and live experience of the extraction of fluids produced a range of emotional reactions in both the spectators and participants.

For Boyle, the response from the public was very positive: "Everyone that was there seemed to find the experience very moving. The dirt and the mystique, the secretness and the sacredness were washed away."[49] Boyle and Hills's show had, however, a scatological character that produced strong reactions from audiences, some of which were a long way from finding the actions to be alluring, revealing, or mystical. Norbert Lynton from the *Guardian* described it as "a realism ex machina."[50] A critic from the *Bristol Evening News* considered the light show "a clinical performance" that was "an insult to civilized man, and those who left the theatre in disgust were more to be admired than those who might pretend that it had any association with art."[51] David Thomson from *Studio*

International noted that the artists practiced "an openness of response so lacking in self-consciousness" that the bodily matters were "disturbing less in themselves than in the fact that they were supplied by the presenter of the performance."[52]

Although Boyle and Hills's actions had a clinical dimension by showing how the body operates, they could not be reduced to physiological facts; it was a sensorial and a transgressively corporeal event. A pioneer in the use of heartbeats and body fluids, and in using audio projections for artistic purposes, *Son et Lumière* was an explosion of pulsating lights, colors, sounds, and sensations. This was an energetic bodily experience in which beams of light were used to expand the limits of the body onto a screen through the projection of electrocardiograms and bodily images. This representation was not intended to be a purely scientific anatomical phenomenon; its aim was to reveal, through aesthetic shapes and sounds, the body's production mechanisms and the liveness of the body.

While it is known that the brain intentionally silences the heart so as not to interfere with one's perception of the external world, in Boyle and Hills's light show, the heartbeats were audible, setting the pace of a performance in which the mechanistic vision of the body had a more prominent position than the feminine and emotional qualities of the heart.[53] The mechanical image Boyle and Hills presented of the body and the heart was an alienated view of our needs and functions, and it was this relation that provoked uneasiness in many audience members: to see their bodies, their hearts as mere clinical machines of observation. The performance challenged the audience members to consider their preconceptions and societal conventions of the human body. According to Boyle and Hills, "Nothing is more radical than the facts."[54]

ELECTRO-TRANCES: DIANA DOMINGUES

A steady pulse of drums, heartbeats, and shaman chants lures you into a dark gallery where you can experience a voyage inside your own body. In contrast to Boyle and Hills's light show, which blended lights, drums, and bodily rhythms to depict the body as a performing instrument, Diana Domingues's *Trans-E: My Body, My Blood* (1997) proposes a mestizo-electro-corporeal-ritualistic-immersive experience. In this hybrid rite, the electrical energy of the body and the energy of technologies interact and are intended to offer a spiritual experience that marries technology and ritual, multiple realities, and Western and non-hegemonic epistemologies.

Born in 1947 in Paim Filho, a municipality in the state of Rio Grande do Sul, located in the southernmost part of Brazil, artist-engineer Diana Domingues is a pioneer in technology-based art in her motherland and in Latin America. Needless to say, science and technology were two areas hardly considered as women's domains. For over five decades, she has been incorporating heartbeats into her work, using them as a tool to create rituals that make us aware of life in its myriad manifestations. Her works incorporate a wide range of cutting-edge technologies, including neon lights, medical images, interactive installations, and immersive VR caverns. Domingues's work has progressed from light and projection pieces to sophisticated immersive interactive installations that blend cultures and energy forms and are intended to produce electro-trances or intense dreamlike states.

Domingues's works set out to enact a ceremony of connection in which a series of energy interchanges and dialogues connect the natural electrical energy of the body and the artificial energy of technologies. Her intention is to produce "virtual hallucinations."[55] For her, rituals are a series of social or personal patterns that occur on a regular basis and give life and meaning. Heartbeats in Domingues's pieces set the rhythm of the ritual and deliver an electrical signal that feeds not only our bodies but also technological machinery.

Eletrourbs, Eletrogravuras (1979–1983; figure 3.9) is Domingues's first heartbeat piece, shown at an eponymously titled exhibition at Galeria Sérgio Milliet in Rio de Janeiro in 1982. In the center of the gallery, visitors sat comfortably in a Thonet-style rocking chair and listened to their heartbeats amplified using an electric stethoscope connected to a speaker. Participants could complement this sonic feedback with a visual image of their own cardiac peaks and troughs via a cardioscope.[56] In addition, the artist hung two wave-shaped red neon lights from the ceiling, which served as a visual reminder of the beating heart. On the walls of the gallery, a series of Xerox prints depicted objects that are part of everyday life

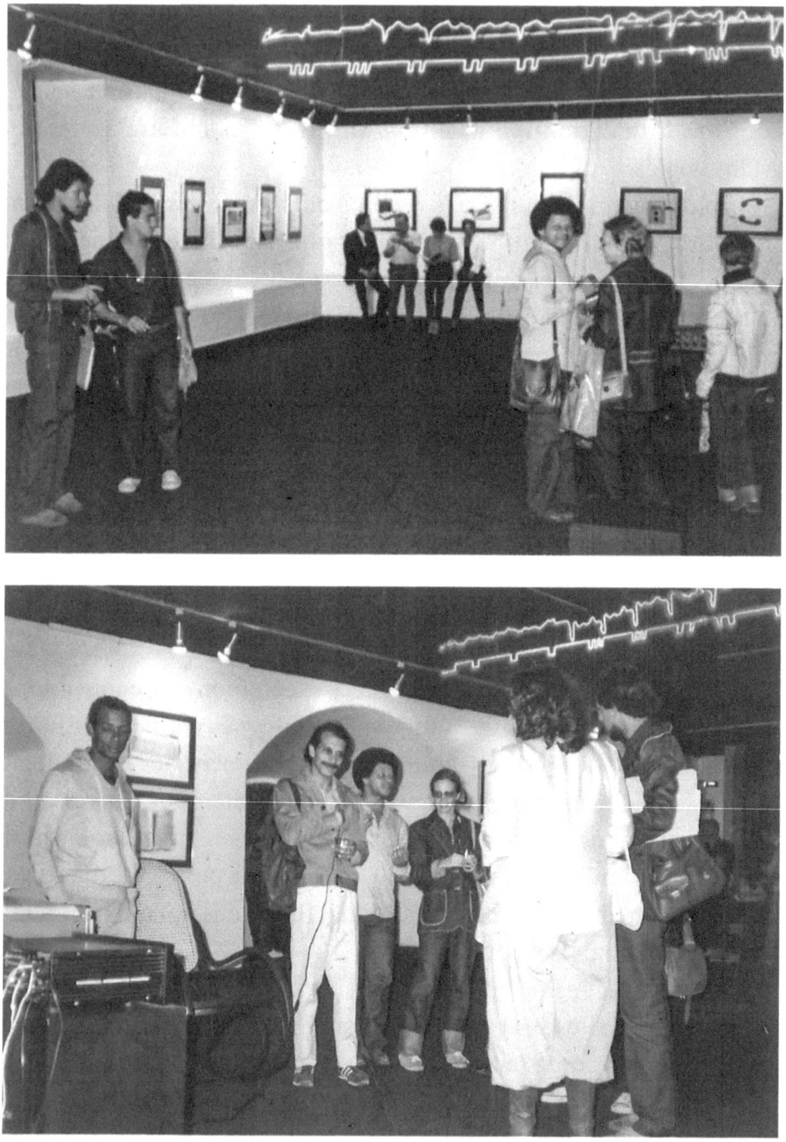

3.9

Diana Domingues, *Eletrourbs, Eletrogravuras*, 1979–1983. Multimedia installation. © Diana Domingues / CNPq, Brazil, courtesy of the artist.

(typewriters, telephones, newspapers) alongside red-colored oscillations. The combination of two forms of inscription, one generated by a machine and the other by the heart, alludes to the symbiotic relationship between the body and technology, which underlies all electronic circuits and machinery in our daily rituals. Most significantly, however, the wavy lines are a reminder that our daily actions are only possible because our hearts beat.

Created a few years later, *Pulsação Ou Onda Ou Vermelho, Não Verde* [In Pulse or Wave or Red, not Green] (1986) consisted of a dark room with an oscillation-shaped neon light accompanied by a strong sound of heartbeats. The project was accompanied by a poem that, like Dupuy's aforementioned *Cone Pyramid*, reveals the bleeding, pulsing and electrical nature of the color red: "Red / Continuous, Discontinuous, / red is electric wave, / power conductor, / red is pulse / Red vibrates. throb. / And flow. / Red indicates life. / Red is vibration. / Red is attention: tension. / Correction / Danger. / Red is noise / Cut. / Red indicates death / It throbs, oscillates, circles, / Activates, vitalizes, life. / Electrifies."[57]

Domingues's heartbeat works from the 1980s were made during a tense political period in Brazil. A repressive military regime seized power in 1964 and maintained control for twenty-one years. Brazilian artists of that time used their art practice as a tool to assert their freedom. They opened the local art scene to international trends, which included the use of medical technologies in the creation of artworks. Brazilian Fluxus artist Paulo Bruscky, for instance, worked in hospital administration for many years, and this gave him access to biomedical tools, including EKGs and EEGs. In *Sentimentos: um poema feito com o coração* [Feelings: a poem from the heart] (1976), he used an electrocardiograph to merge knowledge and emotions, two concepts that are typically conceived as polar opposites.

Diana Domingues herself did not work in the male-led health sector, but she was well-versed in medical technologies:

> My father was a doctor, my husband is a doctor and my daughter is a doctor. I've always been surrounded by medical professionals. So, early in my work, I started developing an aesthetic utilizing X-rays, EKGs, MRIs, etc. My interest in heartbeats, however, is not medical; rather, I am interested in heartbeats as a component of life and rituals. . . . As an artist-engineer, I'm interested in the reengineering of life, the senses, nature, consciousness, and culture. . . . I seek

to be a bridge to make life and its biological systems evident. My intention in this post-biological and post-human era has been to expand natural life.[58]

But what is the posthuman? According to N. Katherine Hayles, it consists of a state of being that transcends the boundaries of human embodiment and consciousness through technological developments, such as VR and AI.[59] This approach suggests that there are no clear demarcations between organic bodies and artificial computer simulations, or between biomedical imagery and the carnal body, or between heartbeat data and the sounds and movements of the heart itself. Domingues's concern with the posthuman and the relationships between the body and technologies is evident in her works from the 1990s when she developed electronic rituals with heartbeats in more intricate, detailed, and intercultural installations.

Trans-E: The Body and Technologies (1994–1995), first shown at Rio de Janeiro's Museu Nacional de Belas Artes, is a complex interactive environment composed of four installations (*Bio-biblio*, *A-fetus*, *In-fluxus*, and *The Supper*). Here the artist creates a journey inside the human body using medical technologies. The experience begins with *Bio-biblio*, a library of life consisting of medical books belonging to the artist's father, whose image is present and vibrating as a beat of life in the exhibition space. References to the pulse can also be observed on a monitor with a vibrant crimson image, as well as a bowl of "boiling blood" installed in the center of the gallery as a life offering that moves in response to the energy of passersby. *A-fetus* is a tunnel comprised of images of a uterine ultrasound scan with a fetus and tomographs that intend to give the impression of walking inside the body. The visitor's image is projected on a television screen at the end of the tunnel, hinting that the person is the newborn. *In-fluxus* (figure 3.10) is located in a third room featuring a large-scale projection of cardiac echographs, as well as an amplification of a continuous heartbeat sound. According to the artist, *In-fluxus* is one of her most powerful pieces "because you are inside the heart, inside its various layers. So, you inhabit this powerful organ thanks to this huge projection that has a very strong heart sound. People couldn't last long inside."[60] Finally, *The Supper* consists of two images of video laparoscopies contained within two rusted iron tanks, surrounded by six whole sheepskins. In front of

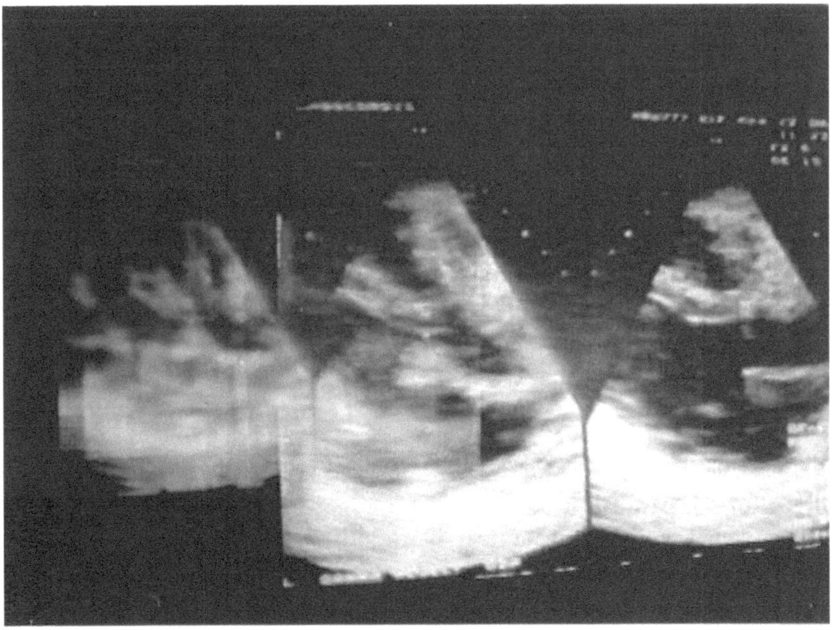

the tanks, a glass-made heart wave hangs from the ceiling dripping red liquid simulating blood.

According to Domingues, the work explores the relationship between the human body and technologies, but in this electronic journey, there are wider issues in play. In *Trans-E: The Body and Technologies*, the biomedical body collides with sacrificed carnal bodies, Western biomedicine with rituality, and thus the vitalist view of the body with the technoscientific

3.10

Diana Domingues, *In-fluxus*, from *Trans-E: The Body and Technologies*, 1994–1995. Multimedia installation with video projector, five transparent screens, electrocardiograms, loudspeaker, sound amplifier, echo device. © Diana Domingues / CNPq, Brazil, courtesy of the artist.

view of the body. Medical imaging in the installation challenges notions of the body as a contained entity while making us aware of how biomedical technologies (such as ultrasounds, echographs, and cardiac recordings) have affected our perception of the body to the point where they can become the body itself. But while Western medicine's approach to medicine conjures biomedical images, indigenous approaches to medical technologies are rooted in indigenous cosmologies and languages, as well as in songs, music, and healing ceremonies.

In her first *Trans-E* installation, Domingues incorporated a bowl of blood, which went on to appear in her later work *Trans-E: My Body, My Blood* (1997–2003). This work and her subsequent pieces were inspired by the time she spent living and learning from the Kuikuro Indigenous people in the Amazon in 1997. Delving deeper into ritual, *My Body, My Blood* is a complex real-time interactive installation that simulates an electronic trance—that is, an altered state of consciousness akin to that experienced by shamans but here induced by digital technology. Blood (an electrically conductive fluid) and heartbeats (as an electromagnetic wave) are an electrical and spiritual energy source that becomes part of a network that creates a "symbiosis of organic/inorganic life."[61] "Electricity is life or life is electricity?"[62] Blood and heartbeats projected outside the body signify life but also sacrifice and exchange.

In rituals, the body is offered during the ceremonies. The idea of sharing the body is common in several rituals of different religions. In Christianity, body and blood are offered to be eaten and drunk. "This is my body, this is my blood" implies an invitation to act as cannibals. The body is also present in other [indigenous] rituals and cults which offer the body and blood of animals. Similarly, in my installation, I ask for the action of the body, and I offer blood . . . which keeps moving inside a bowl that is triggered by the heat of the other bodies captured by the infrared sensor. The bodies of the participants are also swallowed down by the electricity and sent to the machine. The participants don't know that there is a net of electricity connecting the body to the device. . . . I offer the flesh and the blood of the bodies, and I return [heartbeat] sounds and images to them . . . the sound of heartbeats is altered by the actions of the body. . . . I architecturally simulate a cavern to offer a space where people can have "visions" with shamanic powers, as the cavern is one of the spaces where shamans go to meet the spirits.[63]

Trans-E: My Body, My Blood was reincarnated in a more complex version titled *Cavern of Trance* (2007), which included Afro-Brazilian maracas and whistles, as well as images of animals, cave paintings, and lightning in stereoscopic vision to enhance the feeling of being in a trance. Domingues's reference to a cave is significant because the heart is located in the cave of the chest cavity, and, according to several shamanic traditions, the cave of the heart is where one can recognize one's inner nature.[64] Heartbeats, an inner biological phenomenon, are a key part of rituals. First, as I indicated in chapter 1, shamans all over the world, including those in the Amazon and the Xingu Indigenous Park in Brazil, as well as Afro-Brazilian religious communities, employ chanting, drumming, and other sorts of rhythmic noises to help participants attain a meditative state and to connect to the heartbeat of Mother Earth.[65] Heartbeats are also part of ayahuasca rituals because the red-colored brew increases heart rates and blood pressure, which helps in the alteration of perception. Kuikuro rituals are corporeal, spatial, and use repetitive and serial elements.[66] In her installation, Domingues uses repetitive images and the rhythmic noises of heartbeats, drums, maracas, and chants in the same way as shamans to generate a trance.

Domingues's *Trans-E* pieces offer a journey inside the body, and journeys have historically been associated with ritual practices from which people emerge transformed and experienced. Shamans have traditionally been viewed as explorers of consciousness and as healers. Traditional shamanic societies believe in an animistic universe in which everything is interrelated: heartbeats, animals, things, forces. The shamanic trance is a voyage to a nonordinary reality of connection, and it has been demonstrated that during meditation and trance experiences, people's heartbeats beat in tandem, and their frequencies synchronize with rhythmic drumming.[67] This is because percussions (an ancient technology) and hearts (the core of our bodies) speak the same language, one based on vibrations. The boom-boom rhythms of drums, maracas, and sticks communicate with the lub-dub sounds of the heart, stimulating bodily energy, which increases or slows the heart rate and produces dreamlike states.[68] As their rituals involve art (dancing, music, performance, etc.), shamans are both healers and artists who deal with energies. And just as shamans reveal energies in their rituals of ayahuasca in the Amazon and attempt to balance conflicting forces into harmonious order, Domingues's electro rituals also uncover energies and balance forms of knowledge through rituals, heartbeats, and other invisible forces merged with machines. These interactive pieces use electrical feedback to provide multilevel communication and spiritual encounters.[69]

Domingues's installations are what artist Roy Ascott refers to as syncretistic, in which individuals can explore different realities, and computer systems play the role of a shaman in which different levels of consciousness are experienced, as well as different kinds of energy.[70] Through projections, VR, heartbeats, shamanic imagery, and sounds, Domingues intends, as shamans do, to induce an electrical trance in participants, connecting them to nature and to ancient and new perceptions for seeing the world with different eyes. Heartbeats are an electrical phenomenon that, when embedded in rituals and harnessed by technology, have the potential to generate a spiritual experience that can alter our perception of the realm of the real.

In the footsteps of Domingues, other female artists in the late 1990s created autopoietic 3D immersive installations that used electronic circuits and vital signs to generate dreaming-like experiences. Take, for example, Naoko Tosa's *Unconscious Flow* (1999), a biofeedback installation that used 3D computer-generated mermaids to reveal the unconscious feelings of two persons via the measurement of heart rates. The VR pioneer Canadian artist Char Davies created *Osmose* (1995) and *Ephémère* (1998). Both groundbreaking interactive 3D environments offered immersive journeys through natural environments, leveraging the immersant's breathing, balance, and interactive sounds as key navigating tools. *Ephémère* incorporated heartbeat sounds, which had been transformed into ethereal and neutral sounds.[71] Both Domingues and Davies aim to challenge Cartesian dualisms that separate the mind and body, nature and technology, ritual and science, as well as the electrical waves of the body and those of machines.

A later electro-shamanic work by Domingues involving heartbeats is *Heartscapes* (2005; figure 3.11), a VR and augmented reality (AR) journey

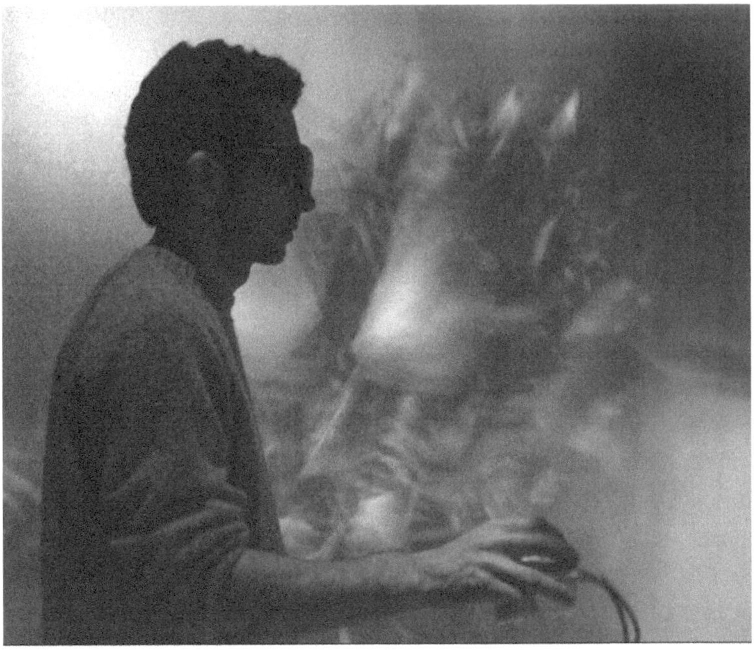

inside the heart, in which she continues to merge heartbeats, noises from indigenous ceremonies, and cosmic sounds. In this electronic ritual, participants can expect to feel embodied and disembodied at the same time since they are within a virtual heart that affects their bodily movements in the physical space, while their electric body data also affects the electric virtual space. Heartbeats (electrical waves) are particularly important in this process since the cardiac frequency detected by the sensors modifies the images presented in real time. The interactivity of the scenes is guided

3.11

Diana Domingues, *Heartscapes*, 2005. NTAV CAVE, immersion in data landscape of the heart controlled by trackers, stereoscopic glasses, biofeedback of heartbeats, and electrooculogram © Diana Domingues/ CNPq, Brazil, courtesy of the artist.

by heart rate variables of 60/80, 80/100, 100/200, 120/140, 140/160.[72] Evoking the dreamlike visions that Amazonian indigenous shamans experience in rituals, voyagers in Domingues's work listen to drums and chants that alter their heart rates while seeing multicolored fluxing shapes, repetitive textures, and animations, as well as the effects of rain, fire, and other natural phenomena. The artist describes this electro-mestizo biofeedback landscape as an "ouroboric perception to emphasize the Latin American mythological figure of the serpent that eats its own tail in a constant autopoietic feedback to the ecocosmos."[73]

In Domingues's heartbeat works, the ontology of life leans toward different forms of electrical energy and levels of reality with the intention of pointing to hybrid or "biocybrid" realities and diverse cultural and medical perceptions.[74] In her work, the artist not only renders divisions between traditional and technological forms of knowledge, and spiritual energy and electrical energy, but she merges them and dissolves them. Like shamanic rituals, Domingues's interactive installations favor processes of becoming over the proclamation of being and propitiate communication between the visible and invisible, nature and culture.[75] Sound, electricity, heartbeats, drums, and moving images are here fluid media, in which technological forces meet vitalist forces, and the Cartesian rationalism of biomedicine and technologies that distinguish between the mind and body, flesh and spirit are contested through a ritual that emphasizes the interconnectedness and pervasiveness of electricity. This form of experience is found inside of us with the potential to create and recreate.

4

TIME

Historically, our heartbeats have been associated with the concept of time because of its temporal and rhythmic nature. It is a sign of time passing, a metronome that counts down the progression of a person's life, reminding us of our mortality. It is no surprise that doctors use heart rates to calculate life expectancy.[1] Time is heartbeats, and heartbeats are time.

Clocks and heartbeats have long been linked; they share a rhythmic and steady beat that is essential to our existence. From ancient times, physicians have been aware that the pulse is the heart's rhythmic language, and they developed ways to count its mysterious and regular movements. These methods were, of course, based on units of time and therefore tied to clocks. From Herophilos's water clock (300 BC) and John Floyer's pulse-watch (1600s) to today's digital pulsometers, the lub-dub of the heart has been related to the ticking of clocks. But the association of heartbeats and clocks extends beyond physical interval measurements. Our bodies are regulated by internal clocks, and biological time is guided by circadian rhythms. This means that the cycles of light and darkness, day and night, interfere with the systole and diastole of the cardiac cycle, as its movements also vary according to our daily activities. Heartbeats, then, are a driving force of our external and internal clocks, of our emotions, of our daily rhythms, of our busy lives driven by the incessant demands of productivity—lives in which the heart may falter but never rest.

Working with both heartbeats and time, some of the artists discussed in this chapter have combined them in order to foster, on the one hand, an awareness of life and, on the other hand, to propose alternative ways of measuring time in our hectic societies. Allan Kaprow, in the 1970s, asked individuals to measure the pulse with watches and recorders in unusual ways in order to make us aware of life, pulse by pulse, counting down. Decades later, using computers and medical technologies, Julieta Aranda created a clock that uses her own heartbeat as a time measure. Her intention was to emphasize the importance of our personal inner time as opposed to any externally imposed time system and notions of productivity. Finally, Sasaki has depicted heart rhythms in discontinuous lines that stretch across time and space, on walls and canvases, marking both personal and collective heartbeats. With these lines, the Japanese artist makes visible the flow of heartbeats during set periods of time, showing us that it is the heartbeat that sustains not only individual people but entire cities.

TIMING PULSES: ALLAN KAPROW

Two men seated at a table next to one another, touching wrists, appear to be concentrating. On the table, a watch, a pen, and a sheet of paper imply the act of recording and counting. Another black-and-white shot in the same document shows the two men palpating each other's wrists, but this time in a stairway. These images are part of Allan Kaprow's *Time Pieces* (1973; figure 4.1), which centers on the monitoring of vital signs and interpersonal relationships. The work was initially conceived as an activity booklet that invited pairs to measure each other's pulse and breathing in more than forty ways.

Born in New Jersey in 1927, Allan Kaprow was inspired by Zen philosophies and created a body of work that involved collaborations, exchanges, the ephemeral, and the conditions of everyday life. His happenings were one-of-a-kind events that required audience participation. While they were meticulously organized, they also allowed for improvisation. *Time Pieces* is part of a series in which Kaprow started to work with smaller groups, producing instruction pieces for individuals and couples. Originally

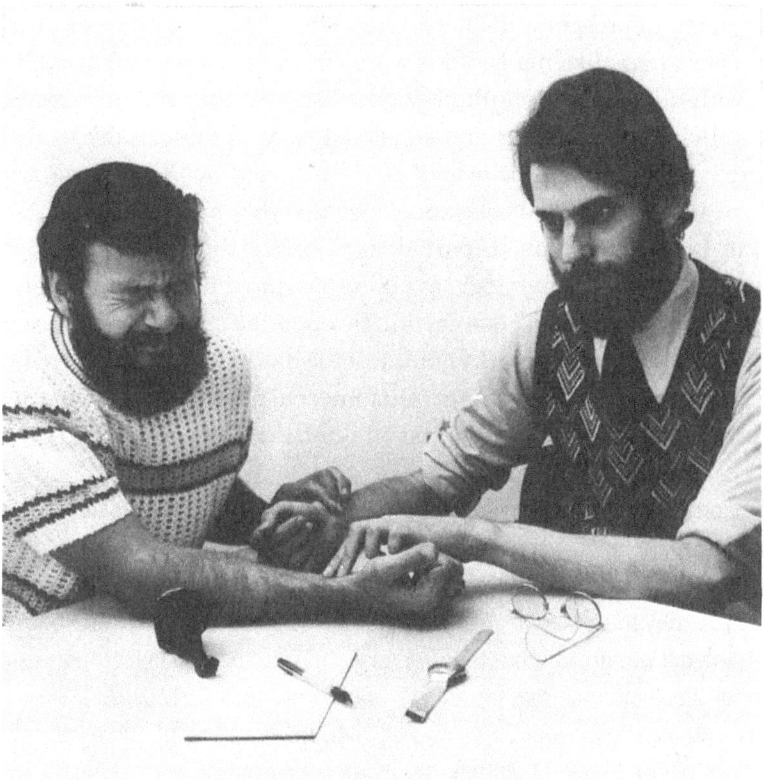

4.1

Allan Kaprow, photo from 1973 activity booklet illustrating how to perform *Time Pieces*, 1973. Presented for *ADA: Aktionen der Avantgarde*, Neuer Berliner Kunstverein, Berlin © Allan Kaprow Estate. Courtesy Hauser & Wirth. Photo by Alvin Comiter.

intended for publication in an edition of the arts magazine *Avalanche*, he later opted to present it in an exhibition at the Neuer Berliner Kunstverein. *Time Pieces*, like many of his works from the 1970s, employed booklets with documentation photographs—staged portraits that indicated the anticipated outcomes and so served as an indexical link to reality. There is also the indexical use of recordings and pulse counting, which are subsequently played back and shown to other participants.

In his presentation to participants before the realization of *Time Pieces* in Berlin in September 1973, Kaprow stated that his intention was to produce an emotional connection through the sharing and awareness of vital signs. Kaprow's work entails interoception, which is the sensing of our heartbeats, breathings, and other internal bodily signals that provide information of our inner state and emotions[2]:

> [The work] concerns human feelings, that is, each participant's feelings. . . . It is a framework of objective moves designed to tap subjective states, states which are, however, unknown at the beginning. . . . These moves are intentionally clinical . . . but the content of the moves are basic to our existence: our heart beat and our breath. Without them we die, and each of us has learned to subtly monitor these internal rhythms as signs of how we feel. . . . Thus, *Time Pieces* can be viewed as a "container" which we fill with the discovery of ourselves. . . . And what we will be doing, of course, throughout the activity is watching ourselves and each other. What we don't know, however, is the effect upon our feelings of conscious involvement in such time segments. This remains to be discovered and reported when we reconvene on Sunday.[3]

Kaprow believed that art should challenge notions of time and space and offered participants a connecting action comparable to the traditional practice of measuring the pulse in the wrist to detect emotions (chapter 1). He eliminated the hierarchical and clinical relationship of patient and doctor, however, by converting the action into a means of communication and relation between people. In *Time Pieces*, Kaprow proposes that time, rather than being considered a marker of events as proposed by Western culture, should be regarded in a more Zen sense, tied to the daily and constant pulse of life, to the time of existence and interconnectedness. With his absurd yet intriguing action, he posits that time is the rhythm of life, the recurrent ticktock of the heart, the regular inhale and exhale sequences of breath. In other words, experiences of life awareness.

Time Pieces involved dozens of participants. They followed more than forty instructions that tested their sense of intimacy and personal space, and implied the consciousness of the pulse and the regulation of breathing. All couples were asked to monitor and tape-record their pulses and then listen to them while apart or climbing stairs. They also sensed each other's pulses while exchanging breaths into each other's mouths as well as captured their breaths in plastic bags and exchanged them while counting their pulses. Kaprow's intention was to generate sensations, interrelationships, and improvisation, but in *Time Pieces*, participants effectively produced a parody of human connection. The activity booklet, with its easy-to-follow instructions and photographic guide to how to execute the work, left little to the imagination. Participants imitated the actions in the pictures, leaving little room for improvisation or self-awareness.

Kaprow later created two separate video versions of *Time Pieces*: one with images of the booklet, a tape recorder, a telephone, and a few pictures of the first participants' pulse and breathing actions, and another with a remake of the instructions with his wife, Vaughan Rachel. Glenn Phillips has noted that "*Time Pieces* are not about watching an activity, on the contrary, they are about experiencing an activity. . . . These books employ a visual system that is so finished and self-contained that the viewer also feels finished after reading the illustrated instructions or watching the video and, thus, does not necessarily feel any urge or need to undertake the activity to truly experience the artwork."[4]

But *Time Pieces* remains illuminating. That is why Kaprow's work is still being performed and recreated by artists today. Take, for example, the re-staging of *Time Pieces* by Tania Brugera's *Cátedra de Arte de Conducta* at the Liverpool Biennale in 2010. These works introduced new ways of perceiving time while emphasizing that heartbeats, time, and art, as components of life, are always subject to changes, recoveries, and rediscoveries. Kaprow's *Time Pieces* offer new possibilities for considering time as something happening right now, with each pulse reaffirming our existence, our ongoing present and presence, and our interconnection with others. It is a living performance that combines the heart and the mind, Western and Asian traditions, the individual with the collective, culture and nature. It enhances the vital but often overlooked aspects of everyday life in a variety of ways; it is an awareness that the pulse is a measurement of time passing that connects us inwardly to our own bodies and those of others. The work is less about the experience of emotions and more about becoming aware of life's sequences and durations that are easily overlooked. By paying attention to pulses and other physical processes, Kaprow reminds us that time and life are meant to be felt and lived rather than known or measured.

HEARTBEAT TIME: JULIETA ARANDA

You are standing in front of a large, white clock that at first glance appears to be a normal device. If you look closely, however, you notice that the clock shows only ten hours rather than the standard twelve and that the second hand moves forward or backward, slowly or swiftly. This fluctuation is governed by the artist's own electrocardiographic data collected over the course of one day. *Two shakes, a tick, and a jiffy* (2009; figure 4.2) is one of many pieces that Julieta Aranda has produced incorporating heartbeats. In all such pieces, the artist resists binaries and the systemic instrumentalization of time.

Born in Mexico City in 1975, Aranda attended the School of Visual Arts and Columbia University, both in New York, in the early 2000s. She currently lives in Berlin. Throughout her career, she has created a body of work that investigates various modalities of time, translation, and circulation. Circulation is present in her constant use of heartbeats that pump blood throughout the body but also in her engagement with flows of energy, materiality, and consumption, including the spreading of art, information, and networking technologies. The latter is evident in her work as editor of e-flux, a platform and archive dedicated to the publishing and distribution of art news and arts writing that she cofounded with Anton Vidokle in 2003.

Two shakes, a tick, and a jiffy was created after the artist visited the Musée des Arts et Métiers in Paris in 2009, where she came across two clocks made in the 1790s during the French Revolution. In 1793, in their bid for a new era and future, the French decided to leave behind the old clock

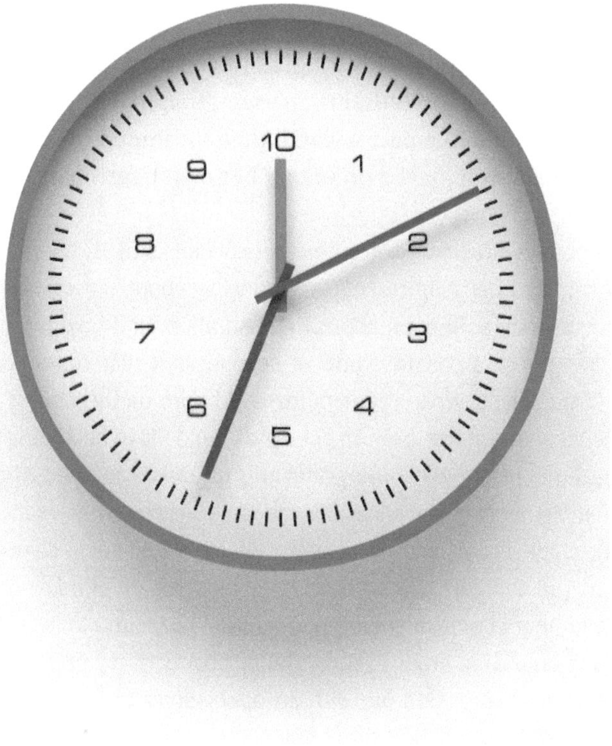

4.2

Julieta Aranda, *Two shakes, a tick, and a jiffy*, 2009. Coated steel, Plexiglas, twenty-four hours of electrocardiography data and operating mechanism, 120.3 × 20.5 cm. Solomon R. Guggenheim Museum / Art Resource. © Julieta Aranda, courtesy of the artist.

system in favor of a new French Revolutionary time. They employed a decimal time system, in which each month was divided into three ten-day periods known as *décades*; each week was of ten days, a day of ten hours, and an hour of one hundred minutes; noon was at five o'clock. The system was not popular and didn't last long; however, it serves as a good example of how time is an illusion rather than a fact. Inspired by this revolutionary act, Aranda decided to create a watch that combined the decimal system with a custom system based on one of her own heartbeats:

> Seeing these beautiful, completely impractical clocks of 10 hours, 100 minutes, and 100 seconds per minute kept me thinking about time as a revolutionary instrument, as well as the concept of recalibration, and how the way we move through time is not a constant flow of seconds but requires a constant recalibration of ourselves. While the minutes and hours on the clock I designed for the Guggenheim move regularly, the seconds hand move according to a 24-hour recording of my heartbeat. During that trip to Paris, I learned I have a severe arrhythmia. My heart beats erratically, as can be seen by the way the seconds pass in the clock piece. Because some minutes have 60 seconds or 120 seconds rather than 100 seconds, the seconds hand must return or advance to allow the minutes and hours to move smoothly. This symbolizes the tension of constantly recalibrating oneself in order to flow. The discovery of my arrythmia, which fortunately is benign, provided me with an opportunity to think about measuring methods, internal clocks, and other vulnerabilities and fragilities. . . . This piece is very important for me since it was the first time I included myself in one of my artworks. And through it, I grasped the thread of what would be my work for the next years.[5]

In addition to *Two shakes, a tick, and a jiffy*, Aranda created many other pieces with her 24-hour cardiological tests performed using a Holter monitor.[6] These include her first ongoing self-portrait, *Slowly/mulled over/matter* (2009–2023; figure 4.3). The work is made up of thirteen sheets of paper with various rows of wavy lines, some of which are visibly discontinuous. *It will, it will. I've guaranteed it* (2014; figure 4.4) consists of a twenty-four-hour audio recording of Aranda's heartbeats and a drawing that translates the recording into binary code. Aranda's arrhythmic heartbeat expands onto the walls of an exhibition space, interrupting not only

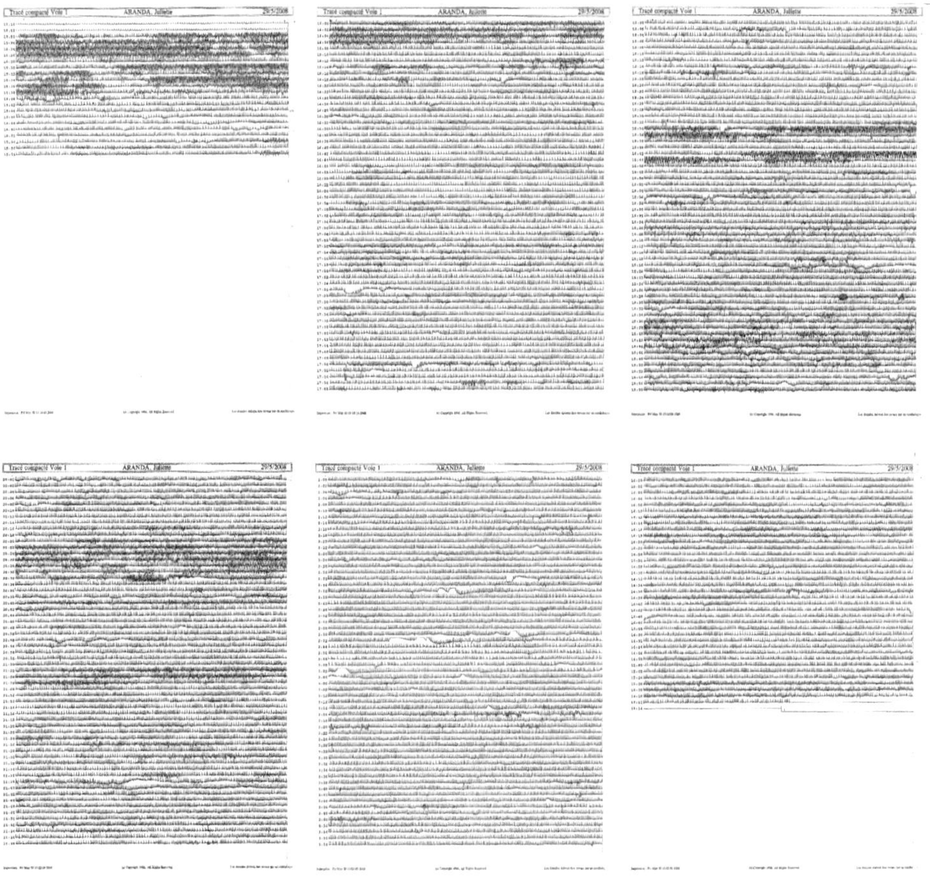

4.3

Julieta Aranda, *Slowly / mulled over / matter*, 2009–2023. Detail, six of thirteen sheets of A4 paper recording twenty-four hours of electrocardiography data. © Julieta Aranda, courtesy of the artist.

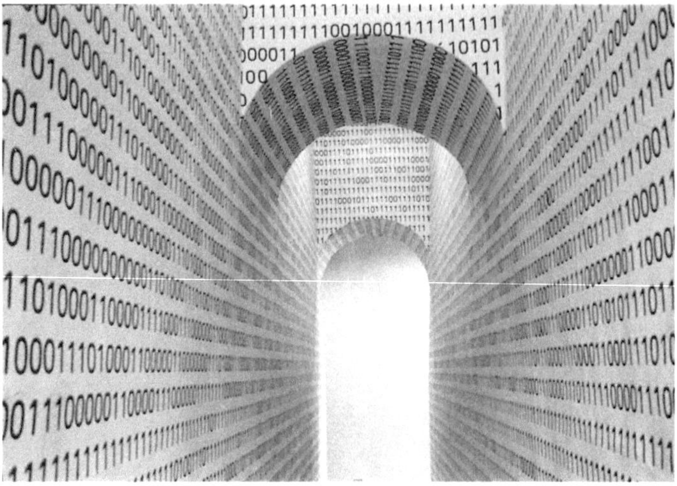

the binary code's rational and continuous stream of ones and zeros but also challenging distinctions between logical and emotional, organic and technological, subject and object, nature and culture.

Aranda's heartbeat portraits question the linear understanding of time that rules our daily lives through clocks, calendars, schedules, and computers. Her preoccupation with temporal politics and an alternative form of time was first articulated in *You Had No Ninth of May* (2008), which entailed a series of installations that map the International Date Line (IDL). Oscillating between the South and North Poles, the IDL at Kiribati is an imaginary graph that traverses the globe so as to delineate the boundary between today and tomorrow. Aranda built a sculpture out of bricks of this line that allowed visitors to consider the rigid yet

4.4

Julieta Aranda, *It will, it will, I've guaranteed it*, 2014. Vinyl print, electrocardiogram data, binary code, audio. © Julieta Aranda, courtesy of the artist.

imaginary systems that regulate our time. Musing on the conception of time as a line, she also made a tape drawing comprising multicolored and multishaped lines that proposed a temporal heterogeneity of parallel moments. The lines' arrangement generates the optical impression of a vibration that announces Aranda's later usage of reverberations as a disruptive time measure.

Aranda's clock is in tune with St. Augustine's conclusion that time is part of our personal consciousness rather than an externally observable and explicable phenomenon.[7] It also aligns with Henri Bergson's observation that time is not a linear and measurable phenomenon but rather a subjective, intuitive, vitalist bodily experience. [8] Following Bergson, Emmanuel Lévinas noted that time is intersubjective, occurring in a moment of suspension of "the diastole and systole of the heart beating dully against the walls of one's skin."[9] In Aranda's work, time is universal and yet defined by personal perception and experience rather than external images and imposed life rhythms. In *Two shakes, a tick, and a jiffy*, the amount of time her clock takes to complete a one-hundred-second revolution fluctuates according to her behavior, emotions, and state of mind. It is no longer a scientific timepiece but a record of the artist's personal experience. Heterogenous forms of time and intersubjectivity are reinforced by the fact that *Two shakes, a tick, and a jiffy* is usually installed alongside *Saving it for later* (2009). The latter work uses a radio transmitter that broadcasts the artist's heartbeat, establishing the sound of time and syntonization as thump-thumps instead of ticktocks.

The connection between time, clocks, and heartbeats is historical: it has been present in the history of both physiology and art for centuries (see chapter 1). Clocks and heartbeats are linked by their different beating sounds, as well as by their function as measuring mechanisms that reflect the pace of life. Both represent life and vitality, as well as the inevitability of time passing. Artists frequently equate clocks and heartbeats with life and as rhythms, means of measurement, and components of our biological clock. Just think on Pablo Picasso's *Science and Charity* (1897), in which a doctor is counting the pulses of a sick woman with his watch. Inspired by her own ailment and created more than a century later with a computer and new medical technologies, in Aranda's work, this relationship is reversed. While in the painting, heartbeats are monitored with clocks, in Aranda's work, heartbeats are the guiding frequency of a clock.

Two shakes, a tick, and a jiffy can be associated with Félix González-Torres's *Untitled Perfect (Lovers)* (1987–1990). This piece consists of two identical clocks installed side by side, touching, representing two lovers, with their "heartbeats" ticking every second in sync. The clocks, mirroring the transient nature of time, love, and mortality, may desynchronize over time because of a dead battery or other misfunction. Aranda, like González-Torres, uses a clock that falls out of sync, but her focus is on the personal difficulty of following an externally forced life pace. Aranda uses a nonconventional watch to question time as a system of control and proposes a new form of measurement: her own heartbeat.

Aranda's proposal to use the heartbeat as a guiding temporal element is more than figurative; the human body literally has an internal clock influenced by rhythmic cycles (heartbeats, respiration, day and night, etc.). Although we have lost touch with our bio-clocks and altered some circadian rhythms by exposing ourselves to artificial light (from bulbs to mobile screens), as well as jetlag and stress, our inner rhythms still guide our perception of time. Recent scientific studies have revealed that time perception is determined by more than just the brain. The heart "plays a fundamental role in our sense of time passing" because the length of each heartbeat can stretch or shorten our time perception.[10] So, as Aranda's clock piece shows, our experience of time is neither linear nor general but influenced by both internal and external factors. Just remember how the pandemic of 2020 demonstrated the malleability of our perception of time and how it fluctuates depending on many social and emotional aspects.

Throughout her work, Aranda emphasizes the varied nature of time and goes on to question time as a biopolitical tool for managing people's heartbeats, bodies, work, and lives. The linear and unilateral time enforced by neoliberal and postliberal societies serves "not only as economic measure in the capitalist system, but also as moral measure."[11] The dominance of time in Western cultures, which is associated with the rise of industrialization, has led to the control of people, territories, and natural resources, and it is intimately related to productivity. Every minute counts in our time-obsessed world because time equals money. As Sarah Sharma has stated, in today's capitalist systems, time needs constant recalibration. This "accounts for the multiple ways individuals and social groups synchronize their body clocks, their sense of future or the present, to an exterior relation, be it another person, a chronometer, an institution, or ideology."[12]

Another work in which Aranda proposes alternative and subjective notions of temporality is *Time/Bank* (2009–). This unorthodox bank platform, cocreated with Anton Vidokle, allows people to trade time and skills instead of using money. It was inspired by alternative forms of currency that emerged during the Great Depression, which led to the time Ithaca Hours, an experimental community currency used in the eponymous city in New York to boost the local economy. Since then, time banking and service exchange have grown into a full-fledged movement, mainly centered on small communities. In *Time/Bank*, participants exchanged activities that were close to their hearts and mutually beneficial. For example, a person who was not allowed to have dogs in their house offered to walk dogs, or a person who loved gardening but didn't have a yard offered gardening services. This notion of time emphasizes interpersonal relationships and exchange rather than clocks, productivity, and money.

For at least a century, other types of time-based currency have been imagined in the history of the arts, some of which have been directly tied to heartbeats. Velimir Khlebnikov, a Russian Futurist poet and writer, declared in many of his writings from the 1920s that heartbeats should be used as a time-based form of currency in exchange for labor. "Effect the exchange of labor and services by means of an exchange of heartbeats. Estimate every task in terms of heartbeats—the monetary unit of the future, in which all individuals are equally wealthy. . . . Use this same unit of exchange to compute international trade."[13]

Nearly a century later, Mexican artist Antonio Vega Macotela also used heartbeats as an alternative currency in *Time Divisa* (2006–2010), a project that explores labor in terms of energy, value, and exchange. In one of the 365 transactions the artist conducted with prisoners in a Mexican jail, an inmate gave Macotela a drawing with a constellation of hearts, one for each pulse he felt on his neck over the course of an hour, in exchange for cooking lunch for one of his relatives. Most recently, French designer Marion Carré created *Payed by heartbeats* (2021), a machine that only accepts heartbeats as payment in order to create unique and personalized compositions for users. It is in the beating of the heart that lies the fuel of labor because without it, no work or task can be done.

Aranda has continued to incorporate heartbeats in her recent work, most notably in *Stealing One's Own Corpse (An Alternative Set of Footholds for an Ascent into the Dark)* (2016–2022). In this film trilogy, the artist continues her investigations on time and the relationship between humans and technology, revealing different forms of colonization, extractivism, and exploitation of life. Musing on space colonization, *Part 1* demonstrates the fragility of a laboratory mouse's heart as it beats in the open during cardiac surgery. *Part 2: Swimming in Rivers of Glue—The Perspective of the Perspective* (2016) employs a soundscape of biological noises, including her own heartbeats recorded during a zero-gravity experience while pregnant. The film is about our experience of space and time. The notion of space extends from outer space exploration to architectural and cityscape reflections and of time as the conqueror of both survival and the future. A work that can be connected to Aranda's films is Chris Marker's *La Jetée* (1962)—a science fiction film on time travel made of still images and permeated with images of animals, technologies, the search of a source of energy, heartbeats and breathing sounds—which asks questions about memory and the future of life on Earth. Aranda uses heartbeats and breathing sounds in her films as synonymous with the inner, outer and biological worlds, the ecology of life, of time passing before our eyes, of Earth's life ticking. *Part 3: Politics Without Oxygen* (2022) shows apocalyptic scenarios of environmental degradation caused by anthropocentrism, extractivism, insatiable ambition, and an ecology of waste. Here, the artist engages with economic time and circulation, including the flow of energy from one form to another, the changes of matter that nature undergoes after being exploited, and mass changes that happen in the production of technological energies.

By suggesting that time is multiple, and heartbeat- and circulatory-based, Aranda's heartbeat works challenge the dominant modern conception of time as objective, homogenous, divisible, linear, and productive. Most importantly, though, she explores alternatives to the current temporal regime characterized by the exploitation of human and more-than-human beings.

THE VITALITY OF SPACES: SASAKI

Lines just one centimeter high stretch from one corner of a building to another. Each is drawn in deep red color as it follows the rhythm of different people's heartbeats. While the heartbeat works I have addressed so far are based on individuals, Sasaki focuses on multitudes. Born in Akita, in 1964, the Japanese artist has made heartbeat drawings his life's activity. He began this "life work" drawn by his fascination with the behavior of vast crowds as he realized that it was literally the heart and heartbeats that make cities alive:

> I was inspired to start drawing heartbeats after traveling to Shanghai in 1995. I saw the old China. Thousands of people were walking on the streets and rushing from one place to another. While I was in the middle of all these people, I could feel their energy and also their heartbeats. I could feel how humans are eager to be alive. Everything starts from the heartbeat. In fact, when we are in our mother's wombs, our hearts are the first organs to form, even before the brain. The heartbeat is the most radical source of humans. This is my perception.[14]

Since then, Sasaki has been painstakingly and obsessively recording his own heartbeats and those of hundreds of other people, marking each one in small wavy lines in an ink of a color he has called "heartbeat red." The artist chose the carmine tone not only for its association with blood but also for its vibrant and striking energy: "Some say it looks scary while others say it is powerful."[15]

Using similar optical effects to Heinz Mack, Sasaki applies his red paint on canvases or in installations guided by his perceptions of sonic vibrations. On some occasions he uses pulse sensors but on others he simply concentrates on feeling his own heart beating. For his durational action paintings, just as Mack did decades before, he wears a silver suit because of its ability to reflect everything around him: "Silver isn't a color, it's just a reflection. If I draw a heartbeat, my mind is almost like a zero, because I have to concentrate on the sound."[16]

The introspective character of heartbeats has inspired many Japanese artists. Take Seiko Mikami as an example.[17] Her various heartbeat works include her renowned interactive installation *World, Membrane and the*

Dismembered Body (1998) in which she amplified the sound of the heart and other bodily rhythms in an anechoic chamber. Another artist is Yoko Ono, whose heartbeat works will be addressed in chapter 7, and Chiharu Shiota, who has included heartbeats in her videos and immersive installations of vein-like webs. What distinguishes Sasaki's portrayals, however, is the twenty years he has spent in the exclusive, repetitive, and persistent use of the vital sign as a drawing subject. The artist documents all the heartbeats that he has drawn; according to the most recent count on his website, this amounts to over twenty-six million heartbeats.[18] His work has an obsessive and serial component reminiscent of On Kawara as it moves from a portrait to an account (as, literally, an accounting). For Sasaki the repetitive and continuous act of drawing heartbeats leads to a meditative state:

> It is like meditation for me too. There are some moments when I am not thinking about anything. There is a sense of openness or joy. Drawing heartbeats is really similar to our lives. There's not much time when I can feel good. Most of the time I feel tired or uncomfortable. But I have a purpose to draw and some physical or mental pain is expected. I draw everyday so my body is used to it just like a marathon runner can run for a long distance.[19]

Sasaki uses the repetitive rhythm of the heartbeat, which he associates with minimalist music, as a meditative act. He believes that it helps him reflect not only on his own life but also on the lives of the many others that surround him. "The meditative state of mind can be the state of nothingness," Sasaki said. "I feel that this state creates 'a pure relationship between the inner and outer world,' and I think that is very meaningful."[20]

As curator Mami Kataoka has noted, although Sasaki's drawings of the heartbeats of thousands of people can sound simplistic, this is far from the case.[21] Within the context of today's society in which wars, pandemics, climate change, and the neoliberal socioeconomic system and its weak health systems in many parts of the world, constantly remind us that simply to "continue living" is a challenge, life itself has become a concern that needs to be recognized, valued, and confronted. Through his art, Sasaki reminds us that the continuous line drawn by our hearts is what sustains our individual and collective existence.

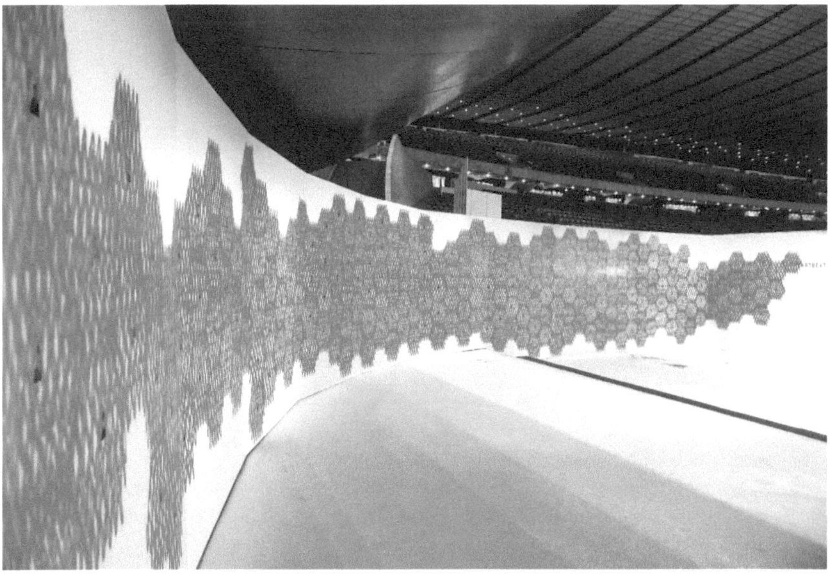

Sasaki's linear durational drawings represent the stream of life that never ceases, flowing from one body to another. With simply a continuous trace of peaks and valleys and the sound of a pen as evidence, Sasaki's paintings depict an unending flow of blood that represents life and time passing. By drawing on the systolic and diastolic movements of the heart, his work is irretrievably related to the temporal, to the organ that gives our lives meaning. All of Sasaki's durational drawings have titles and use processes that engage units of time. For his *In a Day* (2011) series, the artist drew lines throughout one entire day and then partially erased them at midnight. By the gesture of erasing, Sasaki refers to the fact that we are continuously forgetting, not only moments of the day but entire days

4.5

Sasaki, *Connect—Heartbeat Tokyo*, 2011. Performance and installation. Acrylic ink on wall, Polaroid films, sound system, and speaker. Yoyogi National Stadium, Tokyo. © SASAKI, courtesy of the artist.

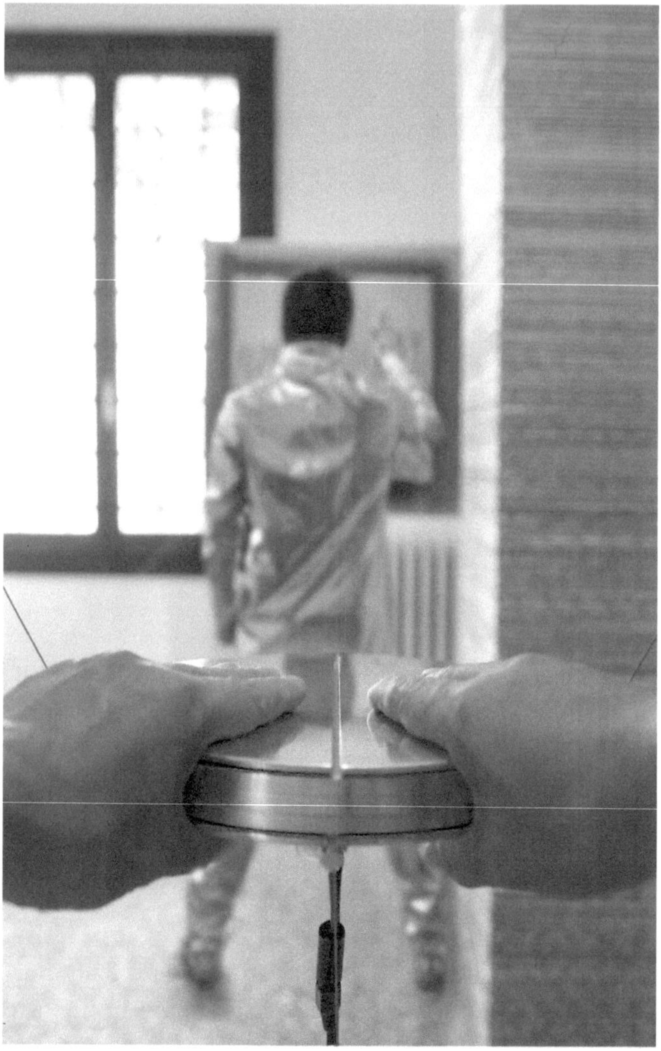

4.6

Sasaki, *Heartbeat Venice*, 2011. Collective drawing and performance. Exhibited in *Personal Structures*, 54th Venice Biennale, 2011. © SASAKI, courtesy of the artist.

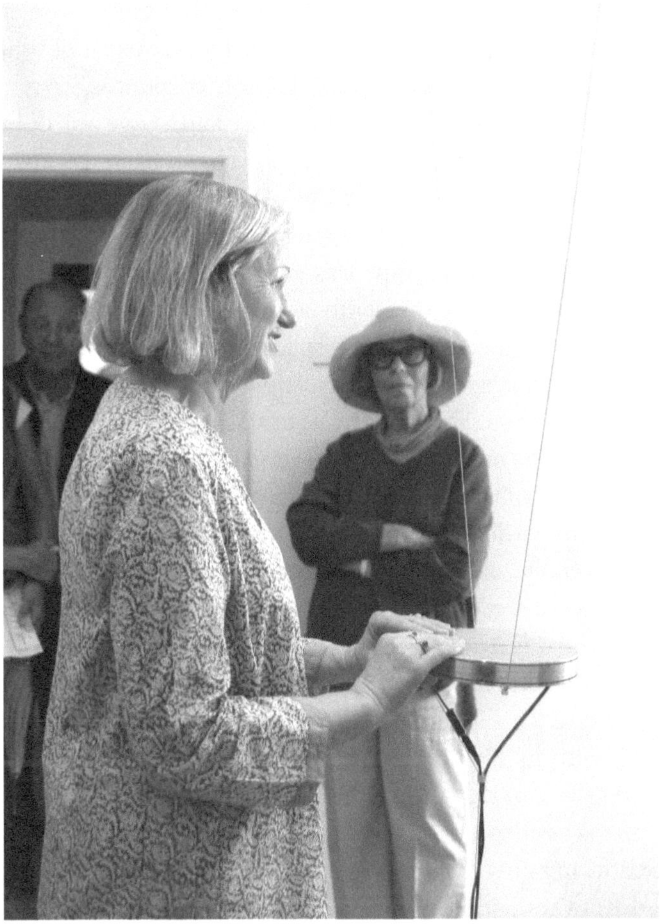

4.7

Participant touching a heartbeat sensor in Sasaki's *Heartbeat Venice* performance, 2011. © SASAKI, courtesy of the artist.

disappear from our memory, as if they never existed. His drawings, on one hand, attempt to capture time; and on the other, are a test of time.

Created during one week each month for three months, *7 Days, August–October 2000* (2000), consisted of a performance and a large wall drawing. For the drawing the artist traced the beating of his heart in consecutive rows of small looping red marks, all within a line of forty-two centimeters wide and four thousand centimeters long that stretched along the walls of the Tokyo Opera City Art Gallery. The stripe was evocative of blood circulation and, like Etienne-Jules Marey's nineteenth-century graphs and experiments on the visualization of blood pressure in space and time, it was a deterritorialization of the interior, a symbolic projection of the circulatory system in a drawing conveyed from the artist's heart to all the exhibition space. Sasaki's hand can be seen as a machine, a stylus extension that translates the rhythm of his heart into graphs. The line here conveys a double meaning referring to volume and to border. The graph is an active volumetric reminder of the incessant movement of a muscle that beats 100,000 times and produces a volume of blood flow of about 2000 gallons per day throughout the body.

In the same way, *15 Days, Heartbeat Drawing* (2006) traced the throbbing of the artist's heart in a circular drawing accompanied by a heart sound recording. This installation, located in the room's center, operates like a heart with its circular shape referencing blood circulation.[22] The time-based aspect of this piece is amplified here by the heart sound recording that is as constant as the ticktock of a clock.

After drawing only his own heartbeat for more than a decade, Sasaki expanded from individual portraiture to social portraiture in 2007. He began drawing other people's heart rates for his series *Heartbeat Portrait* (2007–), which he considers a social art project that aims to make people aware of the heartbeat—that is, the value of life. In 2011, for *Connect–Heartbeat Tokyo* (2011; figure 4.5), he drew hundreds of people's heartbeats as curving red lines that spread across the walls of an exhibition space in Tokyo, resulting in a collaborative artwork. In 2011, at the Palazzo Bembo, for *Heartbeat Venice* (2011), which was part of the Venice biennial exhibition *Personal Structures*, he recorded on separate canvases the beats of 299 visitors to the show, as a way of "communicating one on one." The artist installed a station with a heart rate sensor that reproduced the cardiac signal through a pair of speakers (figures 4.6 and 4.7). The thump-thump of each heartbeat permeated the entire room, while Sasaki marked each one in his distinctive red. At the end of the process, he made the personal portraits into a collective statement by superimposing the images and asking each person to sign the back of their own canvas. "I placed the drawing on top of the others so that person whose heartbeat I drew was the only one who could see his or her own heartbeat," Sasaki said.[23] By suffusing a space with heartbeats, Sasaki makes it apparent that the energy of galleries, museums, and all the other resources of cities are dependent on the invisible movements of the hearts that make sites alive.

Sasaki's drawing practice has been limited in recent years because of problems with his hands, and he has turned to using digital pen tablets.[24] *100,000-beats / 24 hours* (2019) is a monochromatic digital drawing in which a white sphere, made of small and conglomerated lines, floats on a black background. "They are light drawings that can be expanded, contracted, and overlapped, and are flexible," Sasaki says. "Heartbeats have an inherent character in each person and are a commonality that everyone has. I think even the *Heartbeat Drawings*, which are drawn by my heartbeats can be shared with people all over the world."[25] It is not by coincidence that his drawing evokes a planet woven by crossed heartbeats. It is an interconnected environment that reflects the structure of this world. Sasaki uses minimal means for maximum effect. His heartbeats durational drawings are a system of portraits but also a recognition of life. Through these lines, the Japanese artist reminds us that time and blood flow together, always in unison.

5

TRANSPLANTS

The heart has been envisioned by biomedical mechanistic discourses as a sophisticated pump moving 2,000 gallons per day that is not only reparable by surgery but also replaceable via transplants. Transplanting is the act of moving an organ or skin, a system, or a thing into or onto another person or place. The resulting transplant is something uprooted and decontextualized. In this sense, all artworks using heartbeats imply a form of transplantation. They focus on heartbeats and then reinsert them into an object, music, film, performance, or space.

Transplanted life is life beyond death and, therefore, revived life that can go on to inhabit another body. The main focus of the transplanted heart is the heartbeats since they are the sole marker of being in and of the world. Christiaan Barnard's first successful heart transplant in South Africa in 1967 was a media sensation. The *Sunday Times* dedicated a page to the breakthrough, displaying an EKG of Louis Washkansky's new heart to serve as the ultimate "symbol of a medical miracle."[1] When a heart does not beat, it is not a living heart; it is a specimen, an object: white flesh with pipes without blood, blood without circulation; noncirculation implies no oxygen, no moving, no sensing, no thinking, no being, no life. However, the heart's fundamental association with life and death has shifted as hearts can now beat beyond death in another body. So, the heartbeat that declares a person alive is the heart of existence and energy with which an identity can be transplanted. This is because the heart is an organ with multiple symbolic connotations, so it is thought that with the transplantation of the heart, many attributes of a person are transplanted along with it.

Many artists have become fascinated by transplants because they represent ways to explore themes related to life, death, identity, the body, and the relationship between humans and technology. I concentrate here on two artists who have created thought-provoking works that examine the personal, ethical, and societal implications of heart transplantations. Inspired by Dr. Barnard's first transplant, Brian O'Doherty explored the interlacing possibilities of heartbeats, transplants, and identities by creating a series of EKG portraits of Marcel Duchamp. These works symbolically transfer into a series of objects many of Duchamp's attributes and interests. Canadian artist Catherine Richards is known for her studies on the interrelation of the human body and technology, and the hybrid body. She has spent many decades creating works about the electromagnetic activity of the heart and of transplants, and their effects on patients. Many of her works in which she discusses the emotional and physiological impacts of heart transplants are discussed here.

SELF-TRANSPLANTS: BRIAN O'DOHERTY

On the rear wall of a gallery in New York, you find a gray box with a jagged line that throbs and moves up and down rhythmically and hypnotically; this is not just a line, a medical record; it is a person. This heart trace was captured by Brian O'Doherty for his *Portrait of Marcel Duchamp. Lead 1* (1966; figure 5.1). It is not surprising that one of the first artists to use an EKG to represent a subject as a form of transplantation of an identity was also a medical professional. Born in 1928 in Ballaghaderreen, Ireland, O'Doherty graduated as a doctor from University College Dublin in 1952. He practiced medicine in various Irish hospitals while also showing his artwork in Dublin and London galleries. In 1957, he moved to New York to study for a master's degree at the School of Public Health at Harvard University, but his medical career was short-lived. Leaving medicine, he became a television presenter at the Museum of Fine Arts in Boston and then an art critic for the *New York Times*. At the same time, he continued

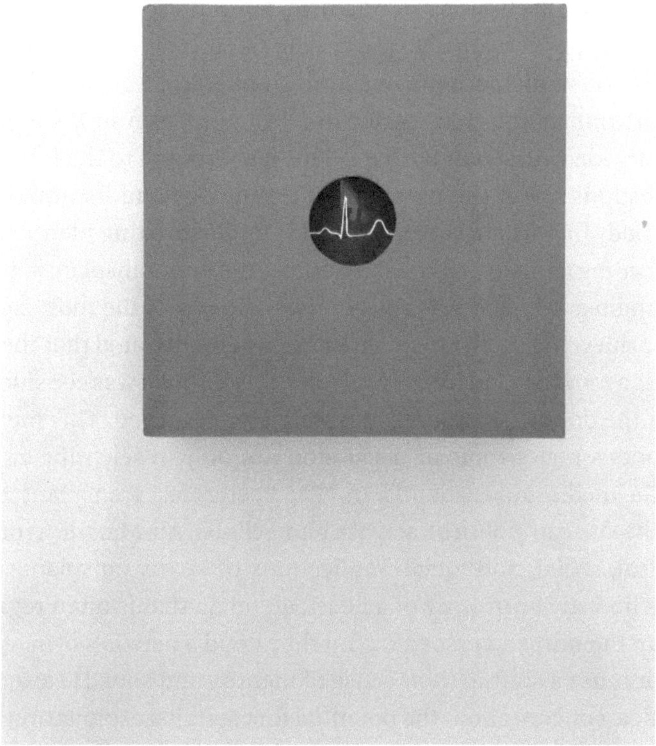

5.1

Brian O'Doherty, *Portrait of Marcel Duchamp, Lead 1*, 1966. Wood, glass, Liquitex, motor, 17 × 17 × 8 in. Collection Hamburger Bahnhof—Nationalgalerie der Gegenwart, Berlin. Courtesy Estate of Brian O'Doherty and Galerie Thomas Fischer, Berlin.

his conceptual art practice, questioning the limits of perception and identity through the use of minimal means, serial sequences, and language.[2]

Brian O'Doherty masterfully combines his interest in medicine and art in *Portrait of Marcel Duchamp*, a series that was at the heart of 1960s art and scientific discussions. This was a time during which the influence of John Cage and Marcel Duchamp expanded notions of art and inspired artists to focus on the everyday, process, and gesture, as well as the use of found objects. Discussions on identity, self-representation, dematerialization, and gender were also at the core of current artistic production.[3] In the field of medicine, a previously unthinkable advance was made: the first heart transplantation, performed in Cape Town in December 1967 by the surgeon Christiaan Barnard. This feat brought to the forefront not only questions about the potential of technology and its impact on the human body but also new paradigms for understanding identity.

According to historian Ayesha Nathoo, the first transplant was like the moon landing, in that it was considered to be one of the most significant human achievements.[4] Transplant surgery demonstrated that the human heart was a movable and interchangeable pump, but it was considered that many of the donor's characteristics were also conveyed. This made heart transplants a major topic of discussion not only in scientific circles but also in the media and academia. In "The Poetics of a Transplanted Heart" (1968), the African political activist and scholar Ali Mazrui wrote about the ethical, social, and racial implications of heart transplants. Mazrui explains how the borrowing of a heart, an organ that is often regarded as central to the human personality, might extend a person's own attributes and even cause a shift in their sense of identity and soul. He also raised a biopolitical concern about the potential for capitalist exploitation of Black people's hearts as spare parts for whites.[5] Transplants, as part of a capitalist system of energy extraction and exploitation, were and are seen to have biopolitical implications for who has access to the procedure and how organs are obtained. Racism and gender bias are present in the history of cardiology, affecting both physiologists and patients. Just consider how African American physician Myra Adele Logan, the first woman to perform open-heart surgery in the 1940s, is rarely mentioned in the history of medicine.[6] The donor of the heart, a white woman named Denise Darvall, is also marginally mentioned in the history of the first transplant.[7] From a medical perspective, the heart is a thing, but for people, it is human. So, Mazrui suggested in his text that transplants could potentially dismantle the flawed concept of "racial mythology—the mythology of blood and its presumed capacity for the purity of identity."[8]

Stimulated by this article, O'Doherty approached the impact of heart transfusions from a very different perspective in "The Politics and Esthetics of Heart Transplants," a lecture given to the University College Dublin Medical Society in 1970. In his text, he describes both Barnard's first successful heart surgery and the notion of identity as fantasy acts:

> The transplanted heart revives, more than anything else, the Frankenstein myth. Yet the artificial heart anticipates a particular science fiction future. The operation is situated between two kinds of imagination—the alienated and the alien, the organic and the inorganic. It points to two kinds of immortality: one in terms of a charnel house of ransacked bodies, the other in terms of technology. These visions, of course, are closely connected to matters of identity and consciousness. . . . For, if I am correct, consciousness, as we know it is the symptom of a lost identity, a paradigm of an obsolete space, and artificiality accepted as fact. Certainly, one can say that the mechanical heart summarizes what I believe is the theme of the age—the swoon of consciousness into technology, the loss of a self that was itself a fiction.[9]

O'Doherty was inspired by the symbolic possibilities of a transference of the sense of identity by the displacement of an organ through the use of science and technology, and by the conception of the self as fiction. On April 6, 1966, he used his medical skills to create a transplant himself. The *Portrait of Marcel Duchamp* (1966–1967) series consists of sixteen pieces, including drawings, objects, and installations. They all center on Duchamp's EKG, which O'Doherty had recorded in his home using a hired electrocardiograph.

> I had hired an electrocardiograph machine. I applied the gel to his wrist and ankles and, as you did in those days, strapped on the metal leads. He lay still, unperturbed. . . . The machine unfurled its long ribbon. The needle was steady, the heartbeat regular. Duchamp stood up—and said, "How am I?" I hadn't given much thought to potential pathology, but if I had remembered my stuff from the

profession I had abandoned years before, I would have noticed what a cardiologist who saw the EKG later pointed out. There (on the EKG) is the PRQST. The Q-R area in lead three and in AVF down here indicate that he had suffered injury to the inferior part of his heart. He finally took in the fact that I was an incompetent cardiologist: "Thank you from the bottom of my heart." And I said, "Thank you for letting me take your portrait." When I recall this, I see what I did not see then: that my desire to capture his heartbeat completely overrode my medical knowledge or interest.[10]

O'Doherty's uses of Duchamp's EKG were multifarious and inventive: *Portrait of Marcel Duchamp: Mounted Cardiogram 4/4/66* (1966; figure 5.2) appropriates the typical electrocardiograph study by presenting it as a framed document. *Study for Second Portrait of Marcel Duchamp: Lead 1* (1967) breaks the beat into fragments and isolates them in multiple circles, making of each a quantum of time. *Portrait of Marcel Duchamp: Lead 1* (1966) consists of a small gray box equipped with a motor that puts Duchamp's cardiogram into motion. With this kinetic work, O'Doherty symbolically extends Duchamp's lifespan to an "infinite continuum."[11]

In O'Doherty's portraits, like that of Étienne-Jules Marey's published in *La Nature* two centuries earlier (shown in chapter 2), cardiograms function as a machine-produced doppelgänger. O'Doherty fondly remembers Duchamp's amazement upon witnessing his own abstract "twin" in motion at an exhibition of *Portrait of Marcel Duchamp: Lead 1*:

I will not forget my excitement when the heart began to beat, inscribing perfectly the course traced on the electrocardiogram by Duchamp's heart. I had him, alive in my hand. . . . I showed the *Portrait of Marcel Duchamp* at the Byron Gallery in my exhibition in 1966 with the drawings and cardiogram. I was in the gallery when Duchamp came. He looked at this record of his own heart beating for some time. Watching Duchamp watching his heartbeat was somewhat eerie. He said nothing. Nor did I. . . . But when [my wife] Barbara Novak met him in the street, he would inquire quite anxiously: "Is it still beating?"[12]

A cardiogram is a trace of an electrical signal, a line created by a recording instrument. Brian O'Doherty's use of a cardiogram, a trace in flux that acts as presence in absence, can be related to Jacques Derrida's concept of the trace. The trace, as understood by Derrida, is "not a presence but is

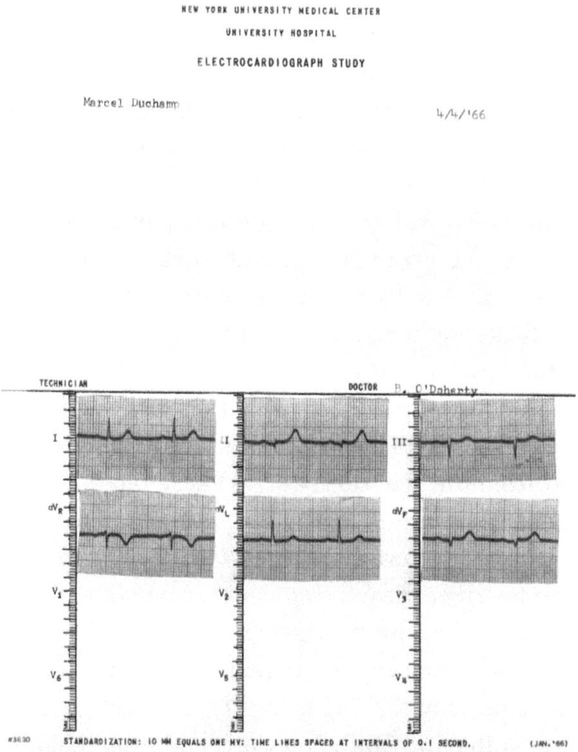

5.2

Brian O'Doherty, *Portrait of Marcel Duchamp: Mounted Cardiogram 4/4/66*, 1966. Print 2012, 28 × 21.7 cm. Courtesy Estate of Brian O'Doherty and Galerie Thomas Fischer, Berlin.

rather the simulacrum of a presence that dislocates, displaces, and refers beyond itself."[13] It is an experience of the past and future through a relationship with the Other, an interplay of presence and absence, inside and outside; it is not an origin but "the origin of the origin." Derrida implies that a trace (and what is a cardiogram but a trace?) is the most minimal form of life when he says, "There is nothing outside the trace."

Portraits, which are a trace of a life, are one of the earliest forms of self-inscription, which evolved from figurative paintings to more abstract forms. To produce a portrait is to provide presence through image and by leaving a trace mark a distance from death. Portrayals aim to capture an identity, a flickering term, but one that can be defined as both the subject and the attributes that constitute this subject. In *Le regard du portrait*, Jean-Luc Nancy points out that portraits are, in a strict sense, "the absolute subject: detached from everything that is not him or herself, removed from any exteriority."[14] In these types of images, Nancy argues, "it is life itself that is recognized and reproduced."[15] If portraits are representations that reveal the essential and most intimate aspect of a person, then heartbeats can be considered one of the most representative and intimate signs for portraying a subject. After all, they are the product of an unseen movement happening beneath the skin that comprises various aspects of a living being. This is because cardiac activity intertwines the mind and body, emotions and soul, making waveforms a representation of vitality.

A cardiograph is like life itself: an apparently continuous line made up of peaks and valleys. Drawn by an organ that is considered both the house of emotions and a pumping machine that puts the body in motion, heart tracings are not simply the representation of a life but of life itself. As each individual has their own specific and unrepeatable rhythmic pattern, heart rhythms are unique to each of us, a feature that makes them a significant symbol, an indexical trace. Graphic traces, as noted by the historian Margaret Iversen, are a hybrid type of representation that "combines the carnal and the symbolic, the line and the flesh, effortlessly, automatically."[16] Cardiograms serve as evidence, therefore, of an existence that can make an individual present even in their absence through this physical connection. In sum, then, they serve as a form of ultimate abstraction. However, these abstract and spiky lines do not *replace* a living person but are instead both sign and signature: a distinctive biological pattern by which someone can be represented or identified. Therefore, heart graphs are a perfect form of abstraction for the portraying of a subject because a cardiogram symbolically contains the essence of a subject's life: its physical, psychical, and emotional force.

While O'Doherty was among the first artists to employ a cardiograph to transplant an identity into a portrait, he was not the first artist to produce a portrait using a body-generated electrical sign. Robert Morris used his brainwaves for *Untitled (Hook, Track, Memory Dents)* (1963), an encephalographic self-portrait recorded at the NYU Medical Center. The image shows eight parallel lines on a sheet of paper, each corresponding to one of the eight different brain regions. Morris's piece, like O'Doherty's cardiogram portrait, is concerned with the mind/body problem: the electrically registered marks are not the thoughts of a person, nor a person, but scientific, objective, and impersonal graphs drawn not by the artist but by a machine.

Although one might perhaps assume that the brain was the most appropriate organ to represent the cerebral Duchamp, O'Doherty focused instead on the heart, an organ that for centuries has jostled with the head for being at the center of the body. O'Doherty's idea of using an organ and its beatings to represent life, the self, and the transfer of an identity has a long history. It's to be found in religious Catholic iconography, in which flaming hearts represent God, as well as in nineteenth-century accounts, such as Edgar Allan Poe's "The Tell-Tale Heart" (1843), in which a person's consciousness is subconsciously transferred into a heartbeat sound. Most significantly, by choosing the line of the heart, O'Doherty meant to emphasize emotion over idea, as well as the warm and enigmatic attributes of Duchamp's personality rather than his thoughts.[17] In the heart, the Irish artist also found an organ that historically has been considered a fetish or relic, alluding to Duchamp's strong influence at the time. Yet O'Doherty's work stops well short of idolization. His motivation for creating the work was his critical interest in refuting Duchamp's claims that "museums were graveyards where works begin to die."[18] O'Doherty considered his work to be the perfect rebuttal to Duchamp's claim by capturing the artist's "electric potentials" and then exhibiting them forever as art in a gallery in the form of "suspended immortality."[19] An article published in *Art News* in 1965, a year before O'Doherty made *Portrait of Marcel Duchamp*, suggested the idea of turning Duchamp into an artwork. "J'accuse Marcel Duchamp," written by *Art News* editor Thomas Hess, criticized the French artist for confusing art with life and for trying "to turn himself into a masterpiece."[20]

With his *Portrait of Marcel Duchamp*, O'Doherty achieved the transplantation of the artist into an art object: "If he [Duchamp] insisted that any artwork on the institutional wall decayed into an antique for the future, that idea now ran full tilt into his own heartbeat."[21] *Duchamp Boxed* (1968; figure 5.3), also part of the portrait series, presents the original record of Duchamp's electrocardiogram rolled up and presented in a small gray cardboard box, painted on the inside with red and blue, two colors commonly associated with the cardiovascular system. The work is a sort of ironic relic in which Duchamp's body—represented by the static record of his throbbing heart—rests placidly inside a box exhibited in a museum.

What is fascinating about O'Doherty's hybrid portrait, where the organic beats inorganically and a cardiogram represents the self, is that the dislocation of Duchamp's heart is a suitable form of representation in tune with Duchamp's identity and aesthetic interests. Using a minimal but significant line, O'Doherty achieved one of the key conditions of portraiture: to expose the intimate nature of a character, their life, and their work. *Portrait of Marcel Duchamp* is in line with the common interest both O'Doherty and Duchamp had in portraiture and identity, and how they used alter egos and adopted multiple personas throughout their careers. Duchamp's *Five-Way Portrait of Marcel Duchamp* (1917), which depicts the artist from various points of view, and *Rrose Sélavy* (1921) (a pun on "Eros c'est la vie"), which presents Duchamp's alter-ego or female persona, are both works in which Duchamp confounds the notion of what a portrait might be and suggests the malleability of the concept of personal identity.

Duchamp's own portraiture practice also experimented with replacing bodies with graphs and objects, as well as recorded invisible processes.

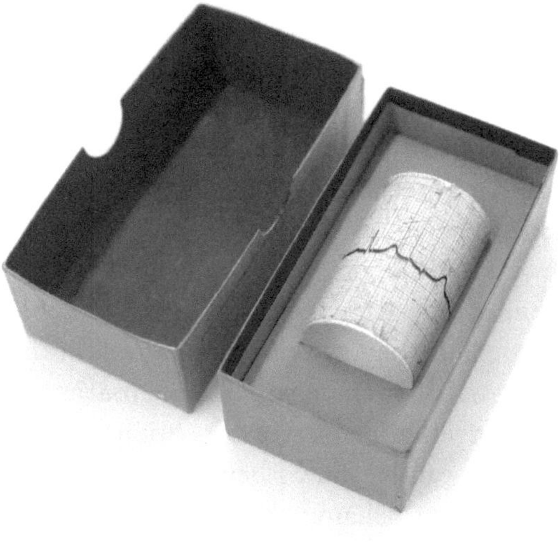

As Linda Dalrymple Henderson has noted, science, technology, and interiority play a central role in some of Duchamp's major works. Between 1912 and 1923, the French artist embedded his work with scientific and mechanical features, transitioning from painter to engineer and finding in the language of mechanical techniques—such as chronophotography and X-rays—a metaphorical substitute for human beings and a way to capture the imperceptible aspects of the body.[22] *The Queen Surrounded by Swift Nudes* (1912), *Nude Descending a Staircase* (1912), and *Bride* (1912) are three examples in which Duchamp was looking to capture the invisible

5.3

Brian O'Doherty, *Duchamp Boxed*, 1968. Electrocardiographic tracing, cardboard box, 3 × 10 × 5.5 cm. Collection Staatliche Graphische Sammlung München, Pinakothek der Moderne. Courtesy Estate of Brian O'Doherty and Galerie Thomas Fischer, Berlin.

matter that hides beneath the clothes and flesh, and energies such as speed and motion. Duchamp's use of lines on top of a figurative painting of nudes in *Network of Stoppages* (1914) is equally significant in the context of his use of mechanical drawings, diagrams, and curves.[23]

O'Doherty's *Portrait of Marcel Duchamp* also brings to the table notions of authorship, which Duchamp had raised in the 1910s by appropriating objects as *readymades*, a concept that remained central to discussions in the arts in the 1960s. Duchamp's readymades transplant every day and mass-produced objects into art. By capturing the line that gave vitality to a body instead of the lines that shaped a body and by using objects in much the same way that Duchamp had, O'Doherty's portrait of Duchamp pays more attention to internal phenomena than to external appearance while also valuing motion and vibration as an essence of life.

Another connection can be drawn between O'Doherty's *Portrait of Marcel Duchamp: Lead 1* and Duchamp's *Anémic Cinéma* (1925). The former uses Duchamp's EKG and the latter *rotoreliefs* or optical disks to create motion graphics that produce the optical illusion of an infinite pulsating motion. The title of Duchamp's film, like the cyclical "lub-dub" of the heart, is a near palindrome, and its reference to a blood deficiency is both a play of words and can be linked to the lacking of color and sound of the film, and also to Duchamp's declared lack of interest in film.[24] *Anémic Cinéma* has been considered by art historian Rosalind Krauss as the first iconic artwork that pulsates.[25] And it is under the spell of Duchamp himself that one of the first works of art visibly pulsates to the rhythm of a person's heart.

According to O'Doherty, while he was recording the electrocardiogram, Duchamp suggested a pun on medical doctor and Marcel Duchamp, and that the work should be signed "Brian O'Doherty M.D.," hinting that his heartbeat was not only an object or material but also the extension of his subjectivity and that therefore he was coauthor of the work. However, O'Doherty declined:

> I wasn't going to allow him co-authorship of his own portrait: his heart had done its work. I have settled electrical potentials describing the functions of his heart, beat by beat, intrinsic to his living, breathing, beating substance, more accurate and unchanging than any representation of the lineaments that change by age of his wonderful face.[26]

O'Doherty, who died in 2022, was the doctor/creator of a portrait that breathed life into inanimate drawings and objects. By using an electrocardiograph, he artistically performed a heart transplant, a displacement of an identity that keeps vitality alive yet is death. O'Doherty's heartbeat portrait moves into metonym territory, where a beat speaks for the whole. *Portrait of Marcel Duchamp* holistically interconnects Duchamp's mind and heart, life and work, capturing and interpreting the totality through a simple but essential rhythm that expands the living self into new territories and forms.

CO-MINGLE: CATHERINE RICHARDS

Your gaze is drawn to two hearts inside bell jars, each charged with plasma gas. Between them is a terrella, which is a wireless electromagnetic model representing Earth that was used to investigate auroras. It contains gases that glow when you and other museum visitors lift the bell jars. This contact excites electrons that in turn make the glow pulsate as if it were connected to each participant's heartbeats.[27] These glass objects become an interface, blurring the limits of the body and providing the virtual sensation of connection between people through corporeal electromagnetic waves. *Charged Hearts* (1997; figure 5.4) explores the possibility of technological communication by participants, which creates the sensory life-like experience of holding and maintaining someone else's heartbeat outside the body.

Born in Ottawa in 1952, Catherine Richards received a bachelor's degree in English literature from Toronto's York University in 1971 and a bachelor's degree in visual arts from the University of Ottawa in 1980. For over three decades, she has been fascinated with the heart and electromagnetic waves and the possibilities for transferring them to another space and using them as a form of communication. Following these interests, Richards has created a body of work in collaboration with scientists and engineers that explores the effects of new technologies on subjectivity and the transgression of bodily boundaries. Her concern with electromagnetic waves and communication led her to heart transplants, but unlike O'Doherty, who was interested in the symbolic transplantation

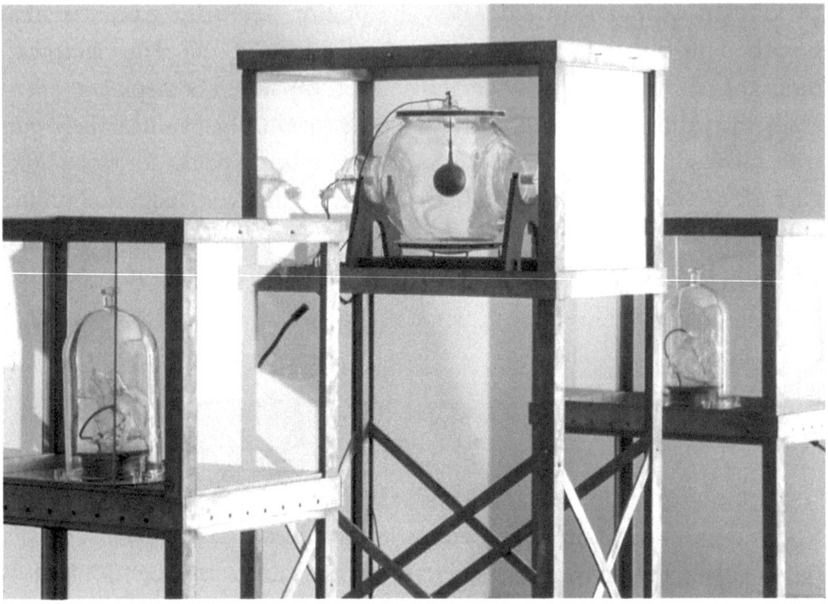

of a person's life through a cardio portrayal, Richards's attention is focused on the physical, cultural, and personal implications of actual heart transplants.

First presented at the National Gallery of Ottawa in 1997, *Charged Hearts* is a multimedia installation that states Richards's pursuit of transplants. The piece mingles the emotional and electromagnetic attributes of the heart, which Richards describes as "the symbolic seat of the emotions," as well as "one of the body's better-known electromagnetic fields. Its tissues are charged with an electromagnetic wave, the heartbeat itself."[28]

5.4

Catherine Richards, *Charged Hearts*, 1997. Interactive electronic installation. © Catherine Richards, courtesy of the artist.

Using old and new technologies, Richards reveals in her work that electromagnetism is not only within us but everywhere and that everything is interconnected. "Our technological culture," Richards says, "has created an environment of electromagnetic signals—pager, television, radio, computer . . . to name a few. As our technology becomes increasingly wireless the space around us is filling up. It is not limited by our body boundaries. We are already cyborgs: part flesh, part machines."[29] The cyborg, according to Donna Haraway, is a late twentieth-century technological postgender creature in which the boundaries between the self and "other," human and nonhuman, automaton and organism, nature and culture, subject and object, fiction and reality are contested.[30] Through transplantations and other biomedical advances, such as pacemakers and cloning, the body has become more plastic, more hybrid, more engineered. As a result, Richards deploys two overarching strategies. On the one hand, she embodies the invisible waves that surround us and that are inside of us by making them visible, and on the other, she explores how transplants might transform our bodies into hybrids and systems of interconnection.

Like physicians during the Enlightenment and the nineteenth century, in *Charged Hearts*, Richards pays special attention to the crafting of her objects and the selection of materials. Her glass hearts are reminiscent of both nineteenth-century medical specimen jars and Marey's artificial hearts models (mentioned in chapter 1), while the terrella is a nineteenth-century artifact.

"The real object in this piece," according to Richards, "is electromagnetic activity and its play between the material and the virtual."[31] A second iteration of *Charged Hearts* includes a virtual component in cyberspace: a game of life, intimacies, death, and survival in which the main challenge is to keep a virtual heart alive by grouping it together with other hearts that, as in real life, can either stimulate or damage your heart. It can be played at home online, but once your heart is critically damaged, it can only be saved by visiting the museum to recharge your heart in the installation. This demonstrates "the artist's own proclivity for the physical over the virtual."[32] In *Charged Hearts*, as noted by Dot Tuer, the heart "becomes a quasi-object, a visual cipher of a field of electromagnetic signals that connects the electrons of our bodies to those of televisions, computers and voltage lines, and connects the cognition of these invisible fields of energy to cultural narratives of intimacy."[33] But, more importantly, I consider that by setting this space of connection, of transferring energy, Richards's work shows that the heart is more than just a pump with electrical qualities that may be transplanted. It is an emotionally charged organ with a range of cultural and emotional implications. In fact, the hearts, which are made of the delicate material of glass, represent the frailty of an organ that can fail physically and can emotionally be broken into pieces. Its health is dependent on affective connections with other people.

A work that relates to Richards's work in terms of its time creation, the transplant and sense of heartthrobs, and the sharing of feelings between strangers through wireless intercommunication is *Mobile Feelings* (2003–2004) by Christa Sommerer and Laurent Mignonneau. Created a few years after Richards's installation, the work consists of two egg-shaped "phones" that are ideally held by two strangers in a room. The mobiles are equipped with biosensors, ventilators, vibrators, and electrochemical systems so that once they are touched, they simulate the pulse, breath, and other bodily signs of the participants. The devices work on a standard mobile network so that they can be used anytime or anywhere. *Mobile Feelings* brings into question the frontiers of the public and the private in our technological societies but goes on to explore the possibilities of transplanting the tactility of the body to other spaces via digital technologies.

Expanding on the heart's emotional connective faculties, technologies, and transplants is Richards's *L'intrus (after Charged Hearts)* (1997–2012; figure 5.5). The installation is the first of a series in which the artist explores the emotional and physiological effects of actual heart transplantations. The installation is a reflection on the effects of damaged hearts that fail to perform their duty and are therefore replaced with that of a deceased person.[34] Installed in the center of a dark room, *L'intrus* is a delicate glass sculpture of a heart in a glass bell jar that is translucent and lifeless until touched by a visitor. Upon contact, the excitement of electrons causes it to glow in the dark, giving the impression that it is a red muscular mass filled with blood that pumps and is therefore alive.

The work takes its title from "L'Intrus," Jean-Luc Nancy's text written after having undergone a heart transplant. In it, he expresses a sensation

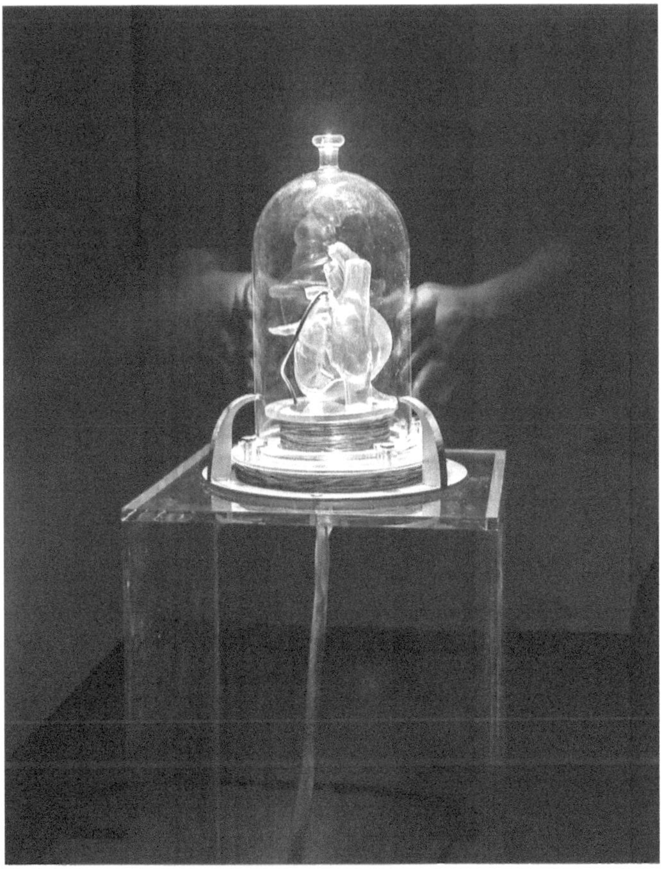

5.5

Catherine Richards, *L'intrus (after Charged Hearts)*, 1997–2012. Interactive electronic installation. © Catherine Richards, courtesy of the artist.

of invasion and an interconnectedness with the unknown and now-dead donor.[35] This is coupled with feeling like Frankenstein's monster, something made up of parts of another body that brings them back to life. "Never has the strangeness of my own identity, which I've nonetheless always found so striking, touched me with such acuity."[36] Nancy wrote, "I am becoming a science-fiction android, or the living-dead."[37] For him, organ transplants "impose . . . the image of a passage through nothingness, of an entry into a space emptied of all property, all intimacy—or, on the contrary, the image of this space intruding in me: of tubes, clamps, sutures, and probes."[38]

L'intrus is a dark and impersonal space with a crystal heart that attempts to transform something familiar into something entirely alien. When visitors hold it, they feel the strangeness of being able to touch and be touched by a disembodied heart. The artificial organ depicts organ transplantation as an oscillation between the natural and the artificial, the living and the dead. Through this pulsating glass object that without human touching remains white and lifeless, the body and the mind, as well as life and death, are symbolically connected, and networks of communication are revealed:

> It is the electromagnetic collusion and collision of our bodies and our new technologies. Here, spectators pick up a heart that "excites" as the spectators themselves become part of this electromagnetic system. . . . This figure, the glass heart, is both hotly fluorescent and ethereal. Earlier we cut open our bodies and drew pictures of it in the light. Now, we use new imaging tools to peer into our imperceptible insides to retrieve images in the dark.[39]

A contrast to Richards's work is *The Body Is a Big Place* (2013) by Australian artist Helen Pynor. This large-scale installation comprises a five-channel video projection, a heart perfusion device, and a live performance. Pynor, like Richards, collaborated with scientists and doctors to produce an installation that evokes nineteenth-century models. Marey's work with animal hearts (mentioned in chapter 1) is clearly referenced, as well as Oscar Langendorff's heart perfusion techniques. Pynor employs the heart perfusion system to reanimate the beating of two fresh pig hearts during a live performance at the installation's site. Here Pynor seeks to create empathy for organ transplant patients by exposing the systolic and diastolic mechanisms that define the limits between life and death and thus the current possibilities of expanding the limits of the heart. By using real, isolated, fleshy, and bloody animal organs, however, she shows the heart as a pump or thing and inevitably brings to mind the vulnerability of other creatures engineered and sacrificed for research.

L'intrus instead reminds us that we are made of chemical, energetic, and symbolic connections that keep us alive. By holding a heart in our hands and by seeing its luminous beat, Richards's installation alludes to the fact that a heart transplant will always be a foreign object that causes a disruption to bodily and identity integrity, and questions where the body begins and ends. In both *L'intrus* and *Charged Hearts*, there is a flow of energy through different systems. The notion of energy refers to electrical activity, electromagnetic waves sourced naturally from the biological realm, translated into different kinds of waves. Pulse-based works such as Richards's involve a total corporal engagement with the artwork that includes the heart, arteries, blood, pulse, hands, skin, eyes, brain, and sound, hence the entire body as a collection of pulsing electrical sensing energy organs. These senses are interconnected thanks to the interactivity of a user who actions the artwork by simply touching a sensor. Following Marshall McLuhan, "tactility is the interplay of the senses, rather than the isolated contact of skin and object."[40] This merging between flesh, organs, and senses filtered through a system permits the artwork to destabilize the boundaries of the body. Further, and more important, by the action of holding a heart, the artist highlights the risks of putting your heart and your life in someone else's hands, whether in the hands of a doctor or in the ability of a foreign organ that must beat to keep you alive.

Transplants, as observed by Jean-Luc Nancy, are both a technical procedure and a "metaphysical adventure." The transplanted heart raises philosophical and spiritual concerns that other transplanted organs, such as a kidney or a lung, do not. This is because the heart is more than a physical organ. Throughout history, the heart has often been conceived as the abode of emotions, thoughts, and the spirit. When it is then removed to another person's body, it is considered that it takes with it something from the owner, that person who has given it. That is why some transplant recipients have reported adopting the donor's personality characteristics.

Scientifically, this effect is assumed to be either a result of cellular memory or the psychological impact of receiving an alien new heart. By an organ transplant, the life of one person is artificially extended, and another person's life from a different background, gender, and age continues in a different body. This life extension raises many ethical and identity concerns. Transplants are part of a capitalist system of energy extraction and exploitation, and have biopolitical implications as to who has access and how organs are obtained. In fact, heart donation is gendered, as more women are organ donors, and most recipients are men, extending the role of women as life givers.[41]

Other technologies like pacemakers, artificial hearts, and left-ventricular assist devices, owing to the scarcity of donors, have been developed, but these technologies cannot replace the unique and sophisticated functions of a heart. Discussions of transplantations inevitably center on the uniqueness of our vital organs and, most importantly, on our interconnectedness as humans, as sensing living beings. The binary distinction between mine and yours, white and colored, young and old, self and other, past and present, natural and artificial, and the living and the dead, are all blurred.

Fear, vulnerability, and anxiety are typically felt by patients on the verge of death owing to an ailing heart. It is not surprising, then, that donated hearts, which are scarce, are frequently regarded as a gift that offers the potential of rebirth.[42] *Still Lives I* (2014), *This Isn't All the Hearts I Have* (2014), *I Can't Let Go of Them* (2014), and *Co-mingling* (2015; figure 5.6) are works by Richards that were inspired by the tokens that patients often give to cardiologists as a thank-you for a successful transplant surgery. The hearts given by patients to doctors can be thought of as votive offerings or *exvotos*. These are small models of an afflicted part of the body (hearts, legs, etc.) that are offered to saints in Catholic churches to give thanks for a healing. Similarly, transplant patients feel gratitude to their donor and to the medical staff that made possible their life continuation and give them objects that demonstrate their appreciation and feelings of indebtedness.

In the cardiologist's office I was startled by small hearts in stone, plastic, glass, fabric. These "knick-knacks" were everywhere. Even after patients recover they

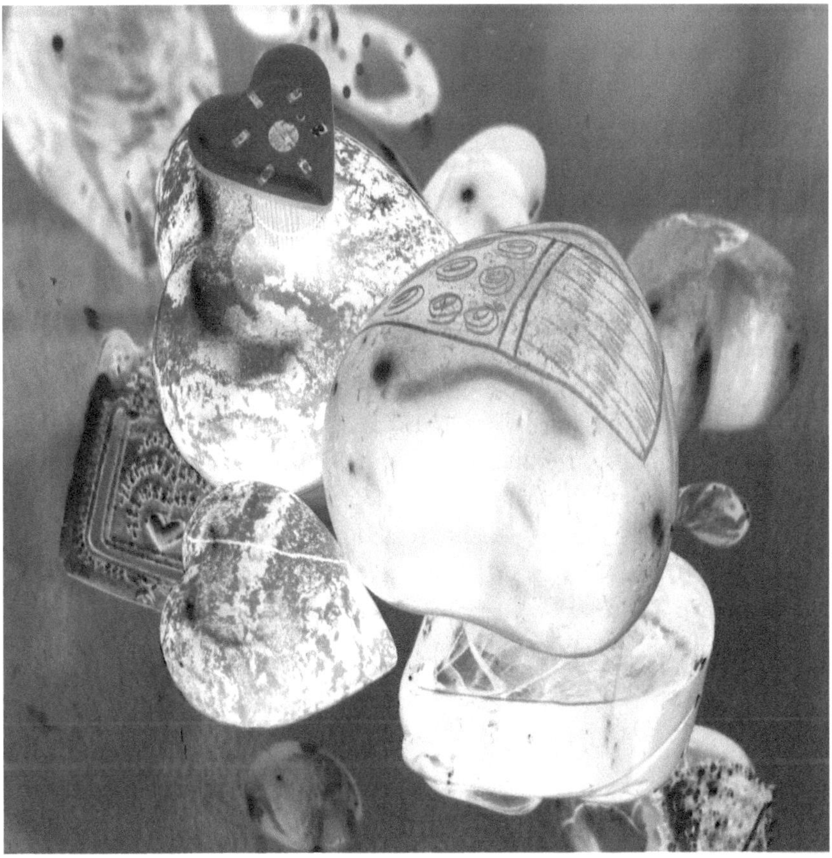

5.6

Catherine Richards, *Co-mingling*, 2015. Stereo passive projection. © Catherine Richards, courtesy of the artist.

repeatedly return to clinic year after year to give her these small tokens. She can identify each one, recall each story and cannot part with them as they pile up on all surfaces. . . . While a living heart is beyond price, these gifts are materially worthless and at first appear inconsequential. But this betrays their power. The patients are setting out to bridge the unfathomable: the new heart was never "dead" but was alive/dead; a stranger died for them to live by having their heart; if their own immune system defends them, the intruder heart dies as will the patient.[43]

Still Lives I (2014) and *Co-mingling* (2015) depict images of floating tokens in positive and negative stereoscopic images that can be watched in 2D or by using 3D lenses. The images are in a space between being alive/not alive, dead/not dead, presence/absence. "I used stereography for these works to point to the power of these objects that is beyond their appearance and to resemble medical imaging: seeing inside something where normally we cannot see," Richards said.[44] These objects of gratitude occupy an ambiguous space that connects two realities and modalities of seeing, as well as two different beings and coexisting elements that beat as one. The heart here oscillates between something symbolic and something material, and that as material it can be observed, molded, exchanged, transported, and inserted. Hearts, and therefore their beats, are an energy that can be harnessed and implanted, becoming both an object of consumption and a gift of life. This makes life and death something fragile, states that are mediated in concert by the domains of medicine, economy, and technology. Transplants are life extended that reflect the complexities of technoscientific advances, biopolitics, human experiences, and the profound interrelation of living beings.

6

SURVEILLANCE

In today's data-driven world, your heartbeat, that vital sign that accompanies you everywhere, is more than just a biomedical sign or a metaphor for life; it can be (mis)used for identification and classification as a control tool. The heart's technoscience is immersed in a complex materiality and socioethical relationship to technological power. This is because cardiac rates, like faces and profiles, have now evolved into records and datasets with dramatically new meanings—they are lines that can tell of a person's inner and outer lives, as well as their biological and physical activities. This gives heartbeats not only meaning but also a sociopolitical dimension, as they have become a tool for the management of life or "biopolitics."

Biometrics, which literally means "life measurement," is the statistical analysis of biological data (heartbeats, fingerprints, voice, or facial patterns) to confirm an individual's identity. Although the use of bodily features for recognition purposes is not new, as evidenced by the use of fingerprints as identifying marks in ancient China (246 BCE), and more than a millennium later in Europe and North America by the Bertillon system, biometrics has grown into a multibillion-dollar industry with the development of computers and recognition systems in the second half of the twentieth century. This growth was accelerated as part of national security measures in the United States and Europe in the twenty-first century, prompted by increased fears of terrorist attacks after 9/11, and rising anxieties about immigration.

Today, biometrics are creating a tyranny of identity by a persistent intrusion into the body. This system is often in service to surveillance by the state as our heartbeats "become inscribed with identities shaped by long standing social and political inequalities."[1] Governments use AI, heartbeats, and other biodata to identify, classify, control, and discriminate against people in the management of their borders, immigration policies, and health-care systems. Private enterprises, such as private health-care providers and banks, have been swift to follow. As many scholars have noted, biometrics operates as a triple-pronged strategy of biologization, informatization, and digitalization.[2] By these processes, identity, movement, citizenship, and access are managed through the automated use and manipulation of our bodies. As a result, these bodies have devolved into informational flows and patterns.[3] So, the ontology of the body, of the heartbeat, relies here on its informational character.

This chapter explores artworks that expose how the heartbeat has evolved into an informational input in our modern era of monitoring and how artists have exposed or counteracted the use of the body as a system of identification and as a subject of surveillance. I start by analyzing the work of Peruvian artist Teresa Burga. She shows us how the heartbeat can be inserted into bureaucratic systems of classification and monitoring by creating a portrait of herself through IDs, cardiac recordings, and other medical data that depict her as information. I then examine the work of British-Palestinian artist Mona Hatoum, which addresses how surveillance has entered into our private lives and even inside our bodies through medical surveillance systems and biomedical technologies. Finally, Mexican-Canadian Rafael Lozano-Hemmer employs biometrics in a variety of artworks to counterpoint the use of biometrics as an alienating and categorization tool by creating aestheticized, connective, and collective experiences.

BODY INFORMATION: TERESA BURGA

Upon entering a gallery, you find a series of large black-and-white photographs of a serious-looking young female, side view and front view, and a series of works on paper lining the walls. All these works are displayed on noticeboards. The mysterious woman in the photographs is Teresa Burga, who has expanded her depiction of herself in *Self-Portrait. Structure. Report. 9.6.1972* from an identity card, mugshots, diagrams, blood tests, cardiograms, and phonocardiograms.

Burga, who was born in 1935 in the port city of Iquitos, is an important figure in Latin American art and a pioneer in technology-based art in Peru. She was part of the group Arte Nuevo (1966–1967), which was instrumental in introducing pop art, op art and happenings to the Peruvian art scene in the 1960s. She studied architecture for two years but abandoned her training to study at the School of Art at the National University of Peru. In 1968, she moved to the United States for two years after obtaining a Fulbright scholarship for postgraduate study at the Art Institute of Chicago. There, influenced by conceptual practices and the arts and technology movement, she started to develop a number of works that engaged with cardiac rhythms that are not only related to systems of representation and dematerialization but to the gradual sociopolitical framing of subjects as objects.

Burga's heartbeat works are all independent but interconnected pieces, with their genesis in *Sin título (Heartbeat Recording and Light)* (1970; figure 6.1). This meticulous drawing in ink on graph paper shows the plans of a sculptural machine on a scale of 1:8. The piece, which demonstrates the artist's knowledge of architectural drawing and her interest in science and mechanics, was created while she was a student in Chicago in the 1970s. At this time, interest in the use of electronic and electromechanical technologies was at its peak, with numerous exhibitions centered on the subject (many of which didn't include women) and with art schools offering classes with a greater technical focus.[4] In an interview I conducted with the artist a few years before her death in 2021, Burga traced the origin of *Sin título (Heartbeat Recording and Light)* to one such program:

> During the Fulbright scholarship, I took some courses on machine assembly, with transistors, circuits, and all that stuff. In that course [in 1970], I drew and assembled a machine that transformed the electricity and sound of my heart and made little light bulbs pulse according to the rhythm of my heart. I materialized that machine a few years later.[5]

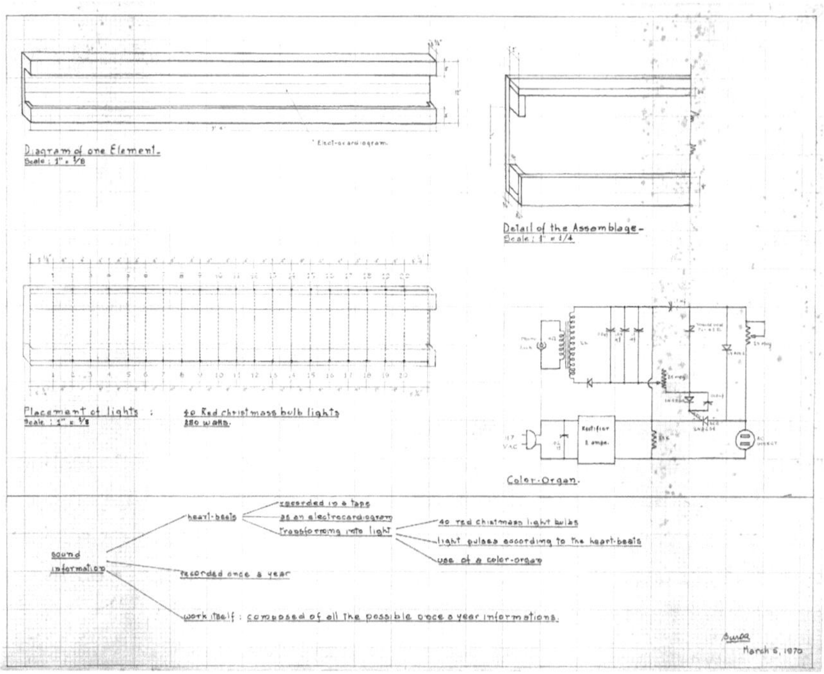

6.1

Teresa Burga, *Sin título (Heartbeat Recording and Light)*, 1970. Courtesy Estate of Teresa Burga and Galerie Barbara Thumm, Berlin.

Heart Beating Machine (1970–1972) is the realization of the machine outlined in *Sin título (Heartbeat Recording and Light)*. This minimalist-looking sculptural piece is an audio-visual representation of the mechanistic condition of the heart and the body as a communications network. Consisting of two white rectangular structures, the interior of the object displays a cardiograph and a series of pulsating red Christmas lights whose blinking is guided by a heartbeat sound. The box becomes a life container in which the ceaseless heartbeats of the artist dwell, extending her presence wherever the piece is displayed. When she returned to Peru, the artist reconstructed her machine and expanded it with other images that animated her self-portrait.

I already had the *Heart Beating Machine*, so I decided to do some blood tests, took some photos of myself, and I did a series of medical studies on a single date to make a representation of myself.[6]

Self-Portrait. Structure. Report. 9.6.1972 (figure 6.2) is one of Burga's most complex works: an anthropometric and interoceptive representation. Comprising medical reports, IDs, photographs, heartbeat sounds, objects, and diagrams that detail the artist's physiological features, it shows a new visualization of the artist. Burga's topological study of herself exposes in detail her physical features and graphic delineations, including the shapes and contours of her face, her bodily structures, the angles of her cardiograms, and many other structural relationships. The documents and objects that comprise the work are her body itself because, as Jean-Luc Nancy notes, the corpus "is immaterial. It's a drawing, a contour, an idea"[7] and "a collection of pieces, bits, members, zones, states, [and] functions."[8] Furthermore, the heartbeat patterns and her other bodily features are the representation of herself or her identity. The piece exposes the intricate relationship between body, identity, materiality, and information.

A hand-drawn diagram introduces and details the three parts comprising the work. "Facial Report" consists of a series of mugshot-like photographs of the artist and facial scans that measure in detail all of her features. "Blood Report" contains a detailed clinical biochemical analysis of the artist's vital fluid. Lastly, "Heart Report" presents the previously

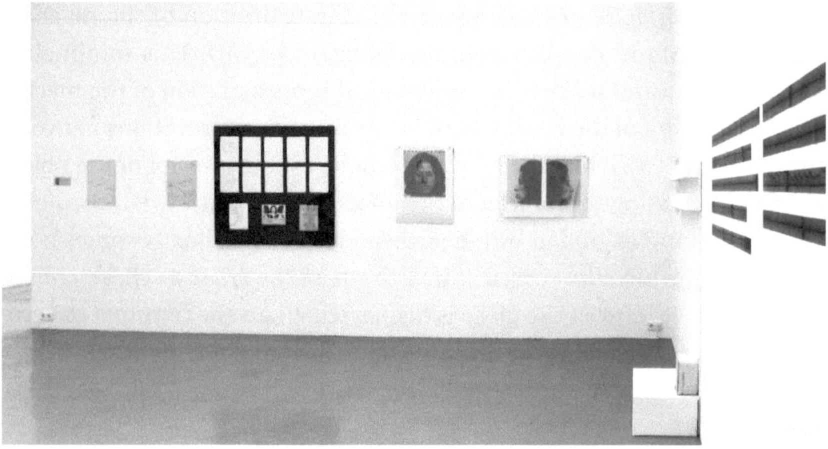

described *Heart Beating Machine* alongside electrocardiograms and phonocardiograms mounted in rows of one or two columns, depending on the space.

Self-Portrait was influenced by Burga's time in the United States, where many social movements were challenging the status quo: student protests were raising awareness of social inequalities through the civil rights movement; there was major resistance to the Vietnam War; demonstrations by feminists for sexual liberation were on the rise.[9] Groundbreaking books, such as *Our Bodies, Ourselves* (1970), empowered women to know and celebrate their bodies and each of their body parts. All these developments together inspired the counterculture and the development of new forms of expression and experimentation. For many artists in this decade, the body became a site for social re-signification. Inspired by these new approaches,

6.2

Teresa Burga, *Self-Portrait. Structure. Report. 9.6.1972*, 1972. Courtesy Estate of Teresa Burga and Galerie Barbara Thumm, Berlin.

as well as conceptual practices and the use of new technologies and science in the arts, Burga developed her own personal conceptual language. Critical but subtle, it did not fit well into the Peruvian art scene that she found upon her return in 1972. The country was now under the authoritarian military government of General Juan Velasco Alvarado, although the regime itself was not a major logistic obstacle to her. During this period, for example, she was still able to access recordings of her heartbeats and other medical data with the assistance of her father's friends, who were on the medical staff at the navy hospital. Such material was not easily accessible at that time, but they were able to give her a detailed analysis of the medical tests that she had undertaken on June 9, 1972, to enable her to capture a representation of herself over the course of a single day. What did become an obstacle for the artist, though, was the local arts community that despised her practice because it was "not Peruvian enough."[10]

Disconnected from traditional media and considered to be under the influence of American trends, Burga's work was contrary to the surreal and expressionist mainstream in Peru and the indigenous nationalistic aesthetics fostered by the regime. Such artistic tastes had replaced the experimental practices of the 1960s in which Burga had played an active role. Nevertheless, in 1972, she did manage to exhibit her heartbeat *Self-Portrait. Structure. Report. 9.6.1972* (figure 6.3) in a two-week eponymously titled exhibition at the gallery of the Instituto Cultural Peruano-Norteamericano (ICPNA) in Lima. At the opening, Burga was subjected to misogynistic and hostile comments that questioned her ability to create such a complex work merging science and technology:

> They thought I was fooling them! But . . . as soon as engineers and some artists saw that indeed it was a real and rigorous topographic rendering, they shut up, because they were about to say that I had falsified it all.[11]

Obscured by her male counterparts and a conservative Peruvian art scene, after that exhibition, Burga's portrait remained in boxes for twenty-five years, waiting for someone to bring it back to life. It was not until the 2000s that Peruvian curators began investigating Burga's art production. After decades of oblivion, in 2010, Burga's heartbeat pieces revived in her first retrospective exhibition, *Teresa Burga. Informes. Esquemas. Intervalos.*

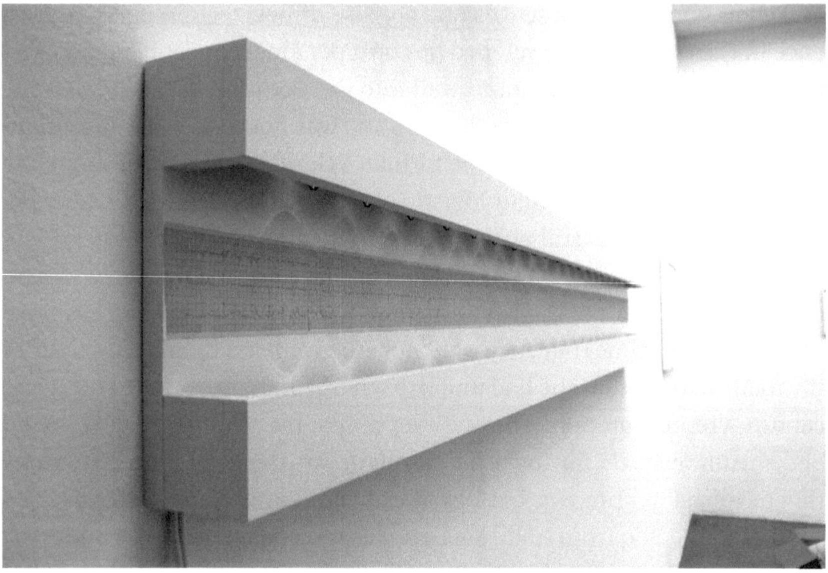

17.9.10, at the ICPNA. This show later traveled to the Württembergischer Kunstverein Stuttgart in Germany, and since then, her oeuvre has begun to be included in major international collections and exhibitions.

While Burga's *Self-Portrait* was very inventive for the time, she was not the only Latin American artist using cardiac graphs and sounds to refer to life cycles or as a form of representation. As mentioned earlier in relation to Diana Domingues, Latin American artists in the 1970s were increasingly drawn to new trends in the arts, and they were starting to experiment with medical technologies. The Chilean artist Juan Downey produced diagrams and schemes of *Inflatable Chairs* (1970) that beat in sync with

6.3

Teresa Burga, *Self-Portrait. Structure. Report. 9.6.1972*, 1972. Detail. Courtesy Estate of Teresa Burga and Galerie Barbara Thumm, Berlin.

the sitters' cardiac and respiratory rhythms. In 1974, the Mexican painter José Luis Cuevas, driven by hypochondria and his fear of death, created a self-portrait using an electrocardiograph. Colombian artist Sandra Llano-Mejía was creating various portrait works with cardiographs like *In Pulso* (1978). In this video, the artist recorded her cardiac oscillations on long sheets of paper. And while European artists like Heinz Mack and Brian O'Doherty had already been using cardiographs as an index and a form of representation, and Boyle and Hills were performing heartbeats and other bodily functions of the body, Burga's *Self-Portrait* broke new ground as one of the earliest and most innovative works not only from Latin America but globally. Her inch-by-inch deconstruction of the body's interiority and exteriority gave a complete topographic render of a person's existence through their biological statistics.

For *Self-Portrait*, Burga's conceptual strategies, such as the use of instructions and diagrams, created an art that intimately related to her life, and, like many feminist conceptualists of that time, she did it using mass-reproducible techniques. These included biological data, photography, text-based media, and Xerox. Relating her art to her working life, Burga appropriated the aesthetics of bureaucracy and exhibited the three physiognomic reports that comprise *Self-Portrait* like the evidence boards of a detective. Burga was very familiar with this style. When she could not find an arts-related job on her return to Lima, she began working for national customs services. She went on to take charge of creating a data classification system that helped manage the customs regulations system by generating reports and charts. Her software was "one of the first successful local implementations for information management systems in the country."[12]

Burga was comfortable with structured analysis as is reflected in her working life, as well as in her artistic life, and in particular in her *Self-Portrait*. Here, data is extracted, classified, and arranged to assist visual analysis. The piece was a reflection not only on identity and information but also on time: on the one hand, it was created during a period when "there was a lot of control,"[13] owing to the political situation both in Peru and elsewhere. On the other, "it was a way of representing the states of the body at a given time," Burga said. "The heart was not going to be the same one beating a year later, my face was going to change."[14]

Burga's diagrammatic project was concurrent with the concerns of the mapping and measurement of the body, which was part of the international discussion around processes of dematerialization within the arts. In the 1960s and 1970s in the United States, as Lucy Lippard describes, "lists, diagrams, measurements, neutral descriptions, and much counting were the most common vehicles for the preoccupation with repetition, the introduction of daily life and work routines."[15] Many artists were using their body measurements as a form of representation. For example, Mel Bochner's *Actual Size (Hand and Face)* (1968) is a photographic work in which the artist explores different ways of self-representation, using measurements of his body parts depicted at full scale. Eleanor Antin's *Carving: A Traditional Sculpture* (1972) also combined pictures and proportions in a grid formed by hundreds of black-and-white photographs of back, front, and side profiles of the artist's naked body. Each registered the day, time, and the artist's weight. Antin, like Burga, objectified her female body, not as a fetish but as information and as a mass to be subjected to critical analysis.

The connection of the female body with medical and biological information was a topical theme in the late 1960s and 1970s. An important example is Martha Rosler's *Bicillin or Medical Treatment II* (1966), a photographic collage of a voluptuous naked woman on an examining table that depicts the female body both as an object and as a subject of medical observation and commercialization. Also, *Vital Statistics of a Citizen, Simply Obtained* (1974–1977), created two years after Burga's *Self-Portrait*, is a performance in which Rosler undresses at the command of medical practitioners who take detailed measurements of her body. Like Burga, Rosler was referring to medical monitoring and the standardization of women in bureaucratic societies.

With its use of mugshots, identity cards, and meticulous physical measurements marked with pen on paper, Burga's piece brings to mind the procedures of Alphonse Bertillon's anthropometries (as discussed in chapter 1) and the later developments made by Argentinian Juan Vucetich in fingerprinting classification. Underlining the connection, Burga's invitation to her 1972 exhibition of *Self-Portrait* was an ID card complete with her photo and personal details. Printed on an IBM 80-column punch card, it directly referenced the serialization of individuals in labor and bureaucratic systems.[16] These mechanisms of examination, administration, and control of people were articulated by philosopher Michel Foucault in 1975, when he asserted they were tools of biopower, imposing bodily discipline and using the body as a document for identification.[17] He observed that human identification scientific studies involved the surveillance and control of the body-self.

As noted by the philosopher Roberto Esposito, years later, this power developed into the "politicization of life and the biologization of politics."[18] Burga's piece demonstrated how the body's ontology had been transformed into informational flows, patterns, and other biocentric data—such as heartbeats, DNA, and facial features. This process blurred any distinctions between the material body and the informational body, transforming them into the same thing.[19] The use of cards and body data prompted the art journal *Correo* to title an interview with the artist "Teresa Burga: An Artist or Computer?" following the opening of her show in 1972.[20]

Besides her *Self-Portrait*, which was created the same year as the seventh and most detailed census in Peru, Burga produced other works related to systems of categorization and control. *Identification* (1978) consisted of a forensic-like text-based piece in which the artist refers to two forms of data evidence of an individual: one direct, through the measurement of the body and the collection of voice tones, finger-, foot- and palm prints; and another indirect, based on eyewitness's descriptions of a person's face, voice, and size. *Profile of the Peruvian Woman* (1980–1981), as its title suggests, presents a complete anthropometric profile (physiognomic, psychological, cultural, emotional, etc.) of the Peruvian woman based on data obtained in surveys conducted on young women.

While Burga's heartbeat work is strongly influenced by conceptual practice, science and technology are at the core of her work:

> [In] *Self-portrait. Structure. Report. 9.6.1972* . . . I specifically asked the doctor to not leave any element out . . . to make a larger and more detailed report . . . Entering a dialogue with medicine and science, affords you something different, particularly to steer away from the mere traditional art. That is what I'm interested in: to interrelate artistic thought and the different sciences.[21]

While *Self-Portrait* explores alternative narratives of self-representation, it is also an identity map and a sort of proto-biometric artwork constructed through the tracing of bioinformation. Burga was responding to the times during which scrutiny of the body was becoming more technologized and more actively sponsored by governments. It was during the second half of the 1960s that computational scientists, like Woodrow Wilson Bledsoe and Mitchell Trauring, were commissioned by intelligence agencies to develop automated machines for facial recognition and fingerprint identification.[22] Face recognition systems, like those used in Burga's "Facial Report," dealt with geometrical anatomical landmarks in the measures of lengths, distances, and ratios of the nose, eyes, ears, and mouth to classify individuals.

Heart recordings and the analysis of blood in Burga's portrayal of herself are largely free of any poetic connotations. They stress instead the use of such body reports as medical data also for the purposes of identification. Blood was commonly used to derive forensic evidence and for identification beginning at the start of the twentieth century with the discovery of different blood types until the 1980s when it was replaced by more accurate systems of DNA recognition.

Burga's "Heart Report" was created months before the first heart transplant was performed in Peru in 1972. At this time, the development of computer technologies, industrializing systems of the classification of individuals, and medicine saw the rise of cardiac monitoring.[23] Burga's portrayal of herself suggests that electrocardiograms can be perceived not only as medical information or as a line that traces a life but also as a pragmatic means of identification, classification, and monitoring. As Donna Haraway noted, information is a "quantifiable element which allows universal translation, and so unhindered instrumental power."[24]

Heart rhythms, along with other body rhythms, have historically been a site of social and political discussions. As Michael Golston has shown in *Rhythm and Race in Modernist Poetry and Science*, between 1890 and 1940, the rhythmic nature of the heart fascinated not only modern writers like Ezra Pound and W. B. Yeats but also European and American scientists in fields ranging from biology, psychology, and eugenics. Some of these specialists, such as Thaddeus Bolton or Christian Ruckmick, were aligned with racist, fascist ideologies and considered rhythm not only as a physical process but also as having social and psychological dimensions that could be put to political use. Some of their studies asserted that heart

rates were determined not by their organic nature but by ethnicity.[25] Following this, non-European populations' heart rates and breaths were used to demonstrate the inferiority of non-Europeans in Peru in the nineteenth century and when eugenics spread in the twentieth century.[26]

Burga's *Self-Portrait. Structure. Report. 9.6.1972* is important within the history of heartbeat works because it is one of the first to use heart rates as an information system, anticipating the use of heartbeats in today's surveillance culture. Burga's work highlights that the body is not only flesh, bones, organs, and physiological processes but also information. Heartbeats in Burga's works offer unintentional signatures that can become the image of a person, an image that can be used for the classification and control of individuals because it reveals not only agents but also their actions. This turns an apparently simple and innocent line from a cardiogram into something that contains a living and sociopolitical subject.

INTRUSIONS: MONA HATOUM

A video mapping the innards of a body, stretching from the esophagus to the colon, is projected on the gallery floor inside a confined space. The visuals, accompanied by the persistent ticking of a heart, make you feel uneasy. *Corps étranger* (1994) is one of Mona Hatoums's best-known works. It connects with Teresa Burga's full-length self-portrait in its use of heart rhythms, medical reports, and information systems, as well as with Mark Boyle and Joan Hills's *Son et Lumiere* in the showing of bodily fluids and functions as a way to make us aware of the inner parts and mechanisms of our bodies. Hatoum's works, however, aim to explicitly highlight how medical systems have intruded inside our bodies, invading our personal and private spaces, including our most intimate orifices and vital organs.

Hatoum, who was born in Beirut in 1952 to Palestinian Christian parents, settled in London after civil war broke out in Lebanon in 1975 while she was on a short visit to Britain. In the 1980s, like many artists of her generation, she started working with performance and later moved to video works that use the human body to confront both cultural conventions and power structures. She uses the human body, focusing on its fragmentations, dislocations, and appearances from a first-person point of view. Her work, as Catherine Grenier has observed, "is like the human body, an indivisible unit inhabited by countless forces, both positive and negative, that mold the visible form."[27] Over her career, Hatoum has created two time-based works that are sonically related in their use of moving images and the sound of heartbeats. More than a century after Laennec's stethoscope sparked anxiety about being used as a spy device, *Look No Body!* (1981; figure 6.4) and *Corps étranger* (1994; figure 6.5) are both works in which the artist disembodies heartbeats and other intimate parts of her anatomy in order to reveal another kind of body: the interiors of surveillance systems, including those of the medical sphere.

Created while still a student at the Slade School of Fine Art in London, *Look No Body!* (1981) is the first work in which Hatoum used heartbeats to explore the body through its fluids and functions. During this forty-minute performance work, Hatoum repeatedly drank water before urinating in a nearby toilet, where she had placed a closed-circuit television camera that was connected to a monitor in the exhibition space. The action was accompanied by a recording that combined the artist's heartbeats, a medical description of the micturition process, and her reasons for creating the work: "It's just that I often wonder where my body ends. . . . I mean what my boundaries are . . . whether it's the skin . . . what about things like hair and nails, and you know, things that come out of the body in the form of urine, faeces, blood . . . where does it actually end?"[28]

These questions as to the limits of the body corresponded with an era when disembodiment and dematerialization in aesthetics and culture were being debated. The mood of the time was captured in Donna J. Haraway's well-known "A Cyborg Manifesto," published a few years after Hatoum's piece in 1985, in which Haraway also questioned the skin as a border of the body. *Look No Body!* was a response to Hatoum's observations of Western culture's estrangement from the body, a shift that the artist considered to have been exacerbated by the 1980s HIV/AIDS epidemic:

> Working with bodily fluids [and heartbeats] was a reaction to this feeling that people around me were so disembodied, they were just like walking intellects that didn't pay any attention to their body and the fact that our body is part of our existence. . . . I wanted to make work that would appeal directly to your senses, that would somehow touch you physically.[29]

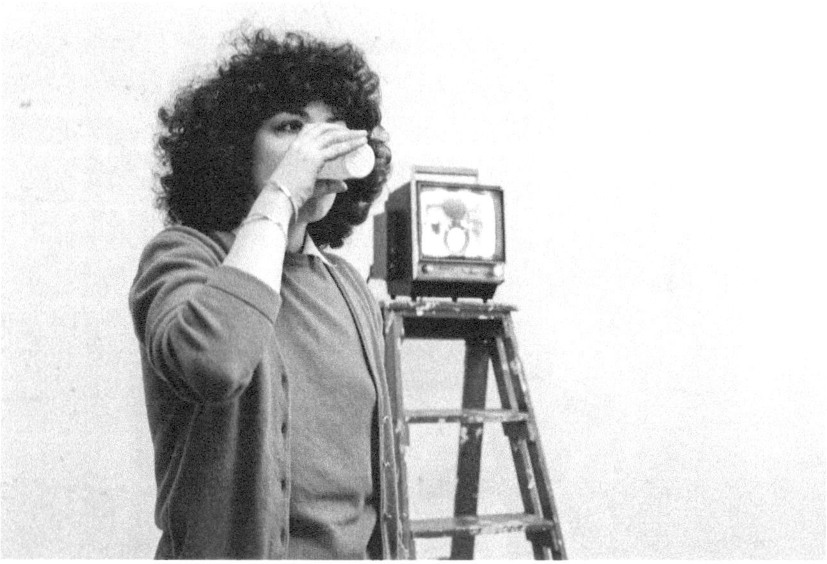

By the disembodying heartbeats and revealing usually concealed private bodily fluids, *Look No Body!* asked audiences to reconsider their bodies: Where does their interior end and their exterior begin? For this purpose, Hatoum used a closed-circuit camera to show us through the walls what is hidden. The device reveals her in an intimate place and moment, urinating in the bathroom. Her intention was to refer to the boundaries being breached by sociopolitical systems of intrusion and control that had begun to spread throughout the United Kingdom and the United States in the 1980s. Thus, the heartbeat sound gives the artist presence in her absence and serves to locate her body in different rooms and situations.

6.4

Mona Hatoum, *Look No Body!*, 1981. Performed at The Basement Gallery, Newcastle-upon-Tyne, 28 March 1981, forty minutes. © Mona Hatoum / Locus+ Archive. Photo by John Kippin.

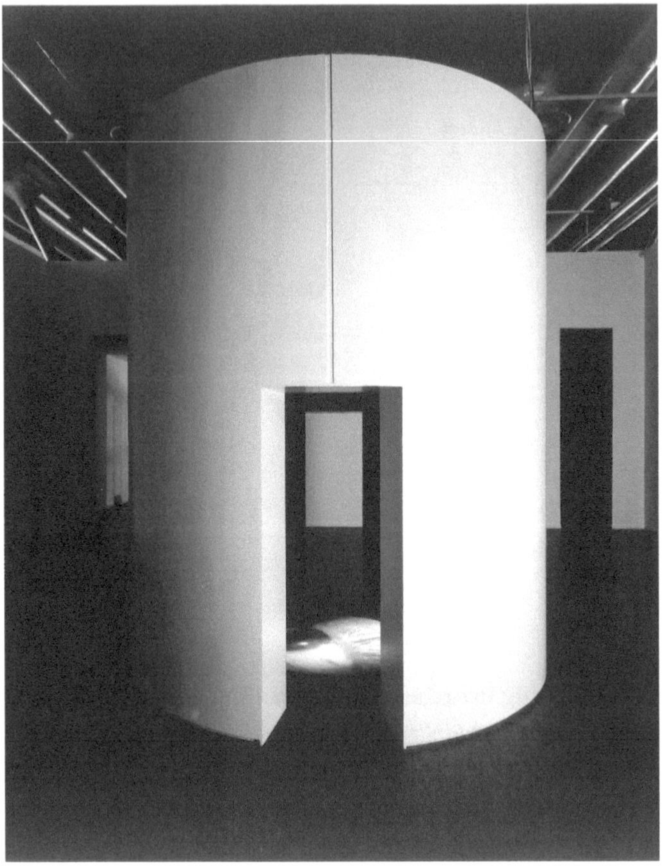

6.5

Mona Hatoum, *Corps étranger*, 1994.

More importantly, though, they are also a reminder of how the heart rate is calm when at rest but noticeably agitated when subjected to situations of stress, shyness, and vigilant external observation.

Heartbeats, medical procedures, and the blurring of boundaries between internal/external, organic/technological, private/public, guarded/intimate are themes that are also present in the works of other artists of Hatoum's generation. Take, for example, Bill Viola's *Science of the Heart* (1982) installation. In a dimly lit room containing a double bed, a pounding and viscous heart being surgically dissected is projected on a screen.[30] The heart, in plain sight, can be linked with life, disease, excitement, or death. The bed, too, is a polysemic object with its deep connections to birth, sex, sleep, illness, and death. Here Viola creates a physical and mental landscape that evokes the various phases of the life cycle through the image and sounds of the heart.

Like Hatoum, Viola uses the video camera and its intrusive lens to set up spaces for confrontation, but Hatoum takes it further. While her works share the peculiarity with Viola of producing in the viewer a sense of trespassing into a very personal space, she puts her own body at stake. There is no place more intimate and vulnerable than everything that is inside our skin, especially for women, whose bodies have historically been objectified, exposed, and condemned at the same time. As Lynda Birke has pointed out, "Self-examination was, is, a radical act in enabling women to see what was hitherto hidden; it thus challenges medical power."[31]

Accompanied by the persistent sound of heartbeats, *Corps étranger* (1994; figure 6.6) is an endoscopic journey to the interior of the artist's body that exposes its organs, tissues, cavities, and inner workings to the viewer's gaze. Consisting of a cylindrical structure that hosts video images shot with an endoscopic camera and projected in a circle on the floor, the video starts with an image of the artist's eye and goes on to enter her mouth, from where the camera initiates a painstaking mapping of Hatoum's internal viscous, fleshy, and pulsating tracts. Recorded with the help of a doctor, using the latest visualization technologies that reveal the inner structures and workings of the body, this invasive procedure traverses the artist's most intimate and secret territories. As Hatoum's heart continues to beat strongly, permeating the dark, enclosed viewing space, anxiety rises in the visitor. "Sometimes people can't take it," the

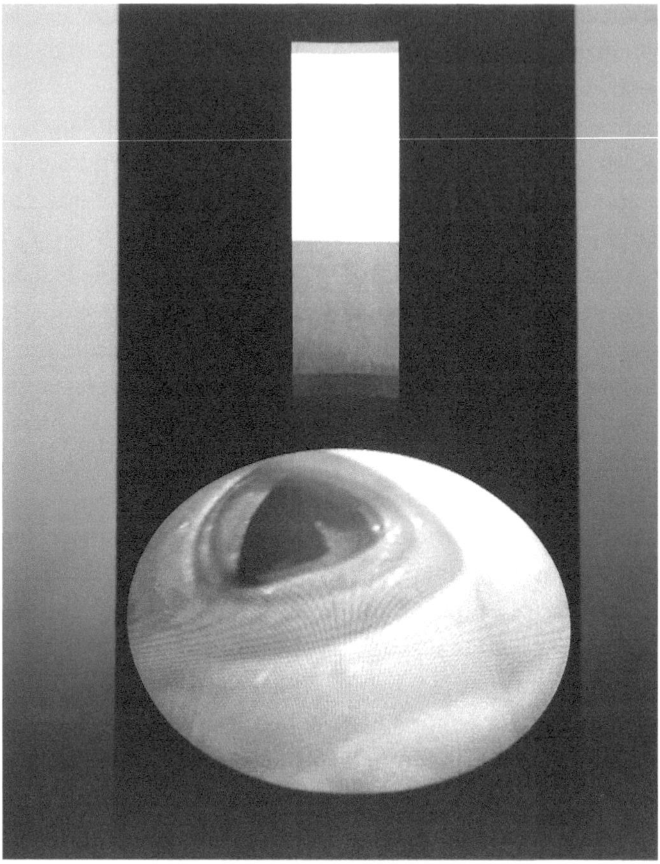

6.6

Mona Hatoum, *Corps étranger*, 1994.

artist remarked, and "it's usually men who are uncomfortable."[32] It is this foreignness with the body that is highlighted in the title of the piece:

I called it *Corps étranger*, which means "foreign body," because the camera is in a sense this alien device introduced from the outside. Also, it is about how we are closest to our body, and yet it is a foreign territory, which could, for instance, be consumed by disease long before we become aware of it. The "foreign body" also refers literally to the body of a foreigner. It is a complex work. It is both fascinating to follow the journey of the camera and quite disturbing. On one hand you have the body of a woman projected onto the floor. You can walk all over it. It's debased, deconstructed, objectified. On the other hand, it's the fearsome body of the woman as constructed by society.[33]

Corps étranger alludes to how women's bodies have always been objects for another's gaze, as well as subjects of control, expectations, scrutiny, and fear (as demonstrated in beat 1). In the piece, the sound of heartbeats makes an unsettling awareness that heightens the (im)material reality not only of Hatoum's body but also of our own bodies. Hatoum explained in an interview how she had obtained the heartbeat recording from the university's teaching hospital opposite her art school:

As a student at the Slade, I thought I was entitled to hassle the doctors at the University College Hospital across the road. They wouldn't do the endoscopy on me, but I managed to get the sound recording of my heartbeat and stomach rumbles, which I used in ***Look No Body!*** And eventually in *Corps étranger*. A piece that I had shelved because nobody would take me seriously. It only became possible to make this work when the Centre Pompidou commissioned me to produce it. Without their influence and financial support, I would never have been able to make it. So *Corps étranger* was a very old idea, but there are a lot of ideas that keep coming in different ways. There are different strands of my work that develop over a long period of time and keep coming in and out of focus.[34]

Ideas also have their own systolic and diastolic cycle and, like blood, can circulate in circles. When *Corps étranger* was finally realized, it culminated in a beautiful, haunting piece. Hatoum made a portrait of her inner structure in order to have a full picture of herself, of her own unknown body, while also displaying the throbbing, humid, and scarlet inside that is shared by all bodies, regardless of age or ethnicity. By presenting bodily images generated with medical technologies, which are usually only presented and interpreted by doctors when we are ill, Hatoum said she "was implying that surveillance can even penetrate inside you. There's no place that remains unturned or unobserved."[35] By turning this "penetrating gaze" to the mysterious body's insides and to our intimate heartbeats, the artist referred to both the medical gaze and the intimidating vigilance of surveillance systems. In sum, to that "all-seeing" force that, as Michel Foucault pointed out, applies both to hospitals and to prison systems.[36] Today, medical surveillance has expanded by the use of body visualization technologies to become a "technobiopower" that, as noted by Haraway, pertains to health and disease, flesh, and film.[37]

Confronting us with the organic and the technological, Hatoum undergoes a medical examination in which she cannot identify herself as a person, as a woman, because viewers see and hear bodily matter and heartbeat sounds. As Judith Butler and Luce Irigaray have demonstrated, "matter" is intricately woven into power relations that surround racism, gender, and sexuality.[38] These heteronormative categorizations are fundamental to surveillance systems. Today, surveillance medicine governs healthy bodies through technology that, as shown in Hatoum's work, projects them as observable and self-observable clinical subjects rather than as persons.

A remarkable feature of Hatoum's work is that it also suggests how "hospital medicine" turned the body into a "three-dimensional object" that could be mapped through heartbeat inspections and other laboratory analyses. Heart monitoring is surveillance equipment for care, control, and management. An echography, endoscopy, and coloscopy are fundamental elements in *Corps étranger*, and they were performed on the artist at her own request. According to David Armstrong, "surveillance medicine," which implies a reorganization and "a fundamental remapping of the spaces of illness," is now dominant.[39] This is because the "medical gaze" has extended from the body out into social spaces and lifestyles.[40] The medical profession has come to consider individuals's daily activities, as well as their cultural and social lives, to be legitimately within the scope of medical intervention. Above all, the focus is on the participation of each person in the "self-surveillance" of their own bodies and health practices. In this bodily discipline, we must take care of our hearts to prevent cardiovascular diseases, the leading cause of death globally—a situation that can be exacerbated by climate change–induced extreme weather, pollution, and fires, all which impose additional stress on our cardiovascular systems. [41]

"Surveillance medicine" addresses each of us through careful observation and data collection by doctors, hospitals, and surveys. As Kate Crawford and Irma van der Ploeg have demonstrated, the widespread use of digital medical records means that medical and body information is monitored by governments and companies that use it to define everything from health insurance rates to medical access.[42] Access to cardiovascular health, for instance, is now determined not only by health providers but also by AI algorithms that assess medical information and can operate with a racial bias, disfavoring women and minority groups.[43] Also, following the current anti-abortion "heartbeat bill" in the United States, heartbeats are also being used as a biopolitical tool to control women's lives and reproductive rights. As we don't know who can access our "digital bodies," the interior of the body is a space vulnerable to control and intrusive observation, as Hatoum clearly anticipated. Furthermore, as noted by Armstrong and depicted in Hatoum's piece, bodies in today's surveillance medicine are as passive as those in a penitentiary system

> The prisoner in the Panopticon and the patient at the end of the stethoscope, both remain silent as the techniques of surveillance sweep over them. They know they are being monitored but they remain unaware of what has been seen or what has been heard [and how it will be used].[44]

The medical and penitentiary systems of vigilance are present in Hatoum's installation. The cylinder cell is a container that makes observers the prisoners of the interior of the body, while viscous images, in combination with the incessant sound of heartbeats, generate sensations of attraction and repulsion to a monitored mysterious body. The cardiac rhythms highlight an awareness of the disconnection between the healthy and the sick, the private and the public, the familiar and the foreign, the visible and the invisible, the inside and the outside in relation to our bodies. With its voyages through the artist's most intimate and secret territories, *Corps étranger* is an unusual, throbbing, and slimy self-portrait that is as intimidating as the intrusions of medical devices and surveillance medicine systems that reveal the status of our familiar but relatively unknown bodies.

BIOMETRICS: RAFAEL LOZANO-HEMMER

An immersive landscape made of thousands of bulbs suspended at different heights creates pleasing oscillating hills that blink to the rhythm of your heart. *Pulse Topology* (2022) is one of Rafael Lozano-Hemmer's most recent biometric works, in which the artist presents a crowd-attracting spectacle of the heart's revolutions as a blinking geography. Born in Mexico City in 1967 and currently based in Montreal, Lozano-Hemmer has created, in collaboration with his Antimodular Research team of engineers and designers, more than ten interactive installations that require the input of heartbeats from participants.[45] Showing how a source has multiple possibilities, all works use a similar interface but achieve very different results. Lozano-Hemmer's work has earned widespread international attention for its spectacular and participatory use of technology.

First presented at the festival *Plataforma* in Puebla, Mexico, and later at the Mexican Pavilion in the Venice Biennale in 2007, *Pulse Room* (2006; figure 6.7) was the first of a series of his biometric works that gave heartbeats the leading role.[46] The installation consisted of hundreds of incandescent light bulbs hung symmetrically by DMX cables at a height of three meters. An interface in a metal structure with two handgrip sensors was located on one side of the space. The sensors were connected wirelessly via a USB plug from a Go!Link adapter to a Mac Mini running Windows XP Pro. This controlled the lights. The artwork is activated when a participant touches the sensors, and the computer senses his or her heartbeats. As visitors grab the interface, their pulse causes one single bulb that is suspended lower than the others, near the handles, to glow. Once the participants loosen their grip on the sensors, there is a blackout in the room, and their biometric rhythms are translated as pulses of light and transmitted to the first bulb in the grid; this displaces the previous

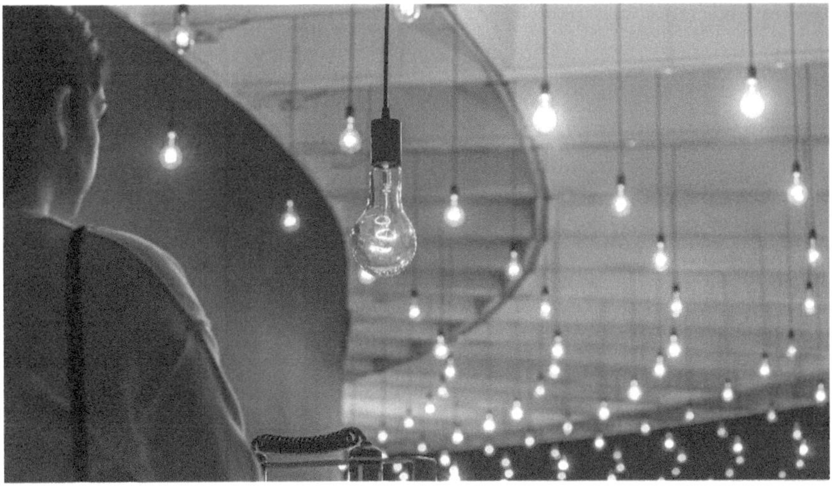

participant's rhythm from that bulb, which is then shifted to the next bulb in the grid, and so on.

The resulting projection of these light rays is a process of reembodiment of the participant's body through vision and pulsation. "We measure 10 variables from the sensor, things like systolic and diastolic activity, and then we convert that into the way the light gets activated. The attack on the tungsten filament is different for each person, so you can see that your heartbeat is quite unique," Lozano-Hemmer explained.[47] The installation, however, dispenses with the notion of the individual since each visitor's rhythms merge into a community of pulses, leaving no trace nor memory. *Pulse Room* evokes a life cycle and serves as a *memento mori* because, just as in life, one existence follows another, pulse

6.7

Rafael Lozano-Hemmer, *Pulse Room*, 2006. Hirshhorn Museum and Sculpture Garden, Washington, DC, United States, 2018. Photo by Miguel Legault. Courtesy of the artist.

after pulse, new registers displacing the old ones. This feeds into a trope about the continuum of life, but it also conveys the inevitability of death and the disappearance of the memory.

The primary inspiration for Lozano-Hemmer's pulse pieces came from a personal experience:

> When my wife was pregnant with twins, I learned that the ultrasound machine can let you hear the heart of the fetus. So, being a nerd, I asked for two ultrasound machines so I could listen simultaneously to the heart of the boy and the heart of the girl. They created this kind of syncopated beat—they would come in and out of sync, kind of like minimalist music. . . . This experience inspired the piece, which shows hundreds of heartbeats all together in a room.[48]

A further source of inspiration was the Mexican film *Macario* (1960; figure 6.8), based on the novel *The Three Guests* by German writer B. Traven and directed by Roberto Gavaldón. *Macario* narrates the story of an indigenous Mexican man who encounters Death and is given the ability to cure the sick. The specific scene that struck Lozano-Hemmer depicts the protagonist who, guided by Death, has a phantasmagoric vision or hallucination in a grotto filled with millions of candles. Each candle represented a human life: the flickering ones stood for people who were alive, the unlit ones for people who had died. The scene presents, on the one hand, the Mexican culture's vision of life where Death's dominant role is evident and, on the other hand, the metaphysics of Judeo-Christian traditions. These place light as the source and origin of the world, leading to the conception of souls as "being-as-light" and life as "being-a-spark."[49] In the film, as well as in Lozano-Hemmer's pulse-based works, the collective nature of life is visualized based on the sharing of space and the understanding of life's ephemerality. Further, the concepts of light and life are clearly related; we see light in the eyes as a sign of life, and in some languages, such as Spanish, the act of giving birth is also called "alumbrar" or "dar a luz," which also means to light up. Furthermore, the *memento mori* in both Lozano-Hemmer's *Pulse Room* and *Pulse Topology* (which recalls Macario's cave) is cheerful, entertaining, and pleasantly accepted as a never-ending succession of light-dark, life-death cycles, and therefore is more related to the Mexican folk vision of Death.

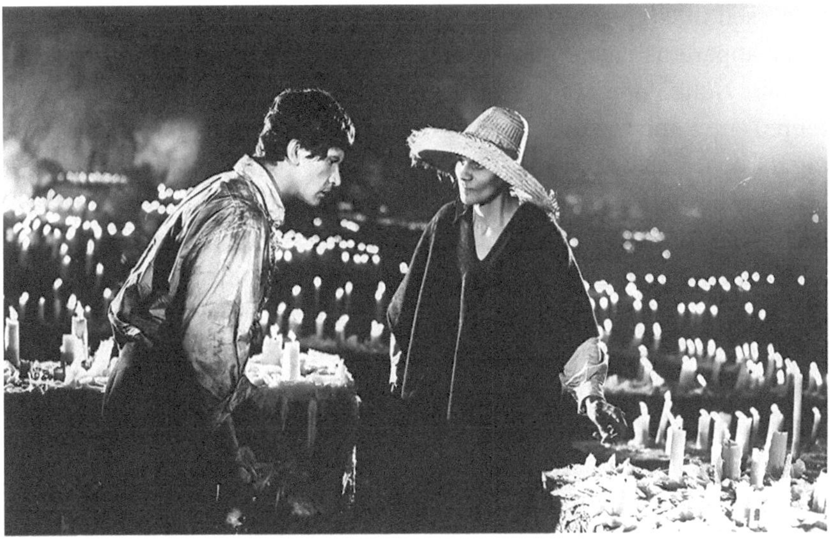

Although Lozano-Hemmer is often considered to be a digital artist working with cutting-edge technologies, he resists interpreting his work as something new:

So for me it's interesting to separate myself from this idea of new media. . . . I really dislike this term because there's nothing new about what I am doing. . . . I find it way more interesting to make connections to the past and see ways in which my work can be related to other experiments that have already been taking place for almost 100 years, than to pretend that what I'm doing is new. . . . My contribution has to do more with the traditions of experimentation.[50]

Lozano-Hemmer's work relates to the history of science and technological experiments in electrical phenomena from the eighteenth century, as

6.8

Still from the film *Macario* (1960), directed by Roberto Gavaldón. Photography by Gabriel Figueroa.

discussed in chapter 1. Particularly relevant to Lozano-Hemmer's work is the apparatus created in the 1740s by the French physicist abbé Jean-Antoine Nollet, one of the most significant physicists of his time. Nollet toured cities and towns across Europe with his apparatus inventions, using darkened salons to show his spectacle of sparks along with electrical body attractions and reactions that kept audiences entertained and impressed. Both Nollet and Lozano-Hemmer created community spectacular performances. In both cases, these performances made electrical phenomena visible via aesthetic devices or instruments linked to bodies. While scientific and technological experiments in electrical phenomena during the eighteenth century opened a space for the culture of performance, spectacle, and education, Lozano-Hemmer's works create a space where technology functions as the interface for the encounter of art, self, spectacle, and collectiveness.

Lozano-Hemmer follows in the footsteps of many artists who have incorporated heartbeats and light. They include Jean Dupuy, and Mark Boyle and Joan Hills, as well as others who have created compelling works. There is Stelarc's *Amplified Body* (1985), a mechanistic performance in which the artist extended his heartbeats and other bodily functions using sensors and light beams. Jonathan Borofsky's *Heartlight* (1991), a tripod lamp that beats and sounds in time with the artist's heartbeat. *Light Blaster: Immaterial Membrane* (1993), by Christian Moeller, which allows visitors to control a membrane of light with the frequency of their heartbeat. Vadim Fishkin's *Lighthouse* (1996) is a work in which the artist live-transmitted his heartbeat to a beacon positioned in the copula of Vienna's Secession Building for many days. Another example is Noriyuki Fujimura's *Heartbeats* (2003), an interactive collective public art performance in which participants literally illuminate the streets with their heartbeats by carrying balloons equipped with bulbs and heart rate sensors. And finally, Christian Boltanski's *Le Coeur* (2005), which is a translation of the artist's heartbeat into a palpitating bulb (see chapter 7).[51]

Although Lozano-Hemmer's heartbeat works have poetic connotations with their references to the finitude of life, the artist is more interested in subverting the use of biometrics as surveillance technology and light as a tool of control or what he calls "violent light." This term encompasses all types of light used for surveillance, interrogation, and forms of human intimidation:

> I am not so much interested in the materiality of light, but the space of light. My interest in light, however, is in the light used by American border choppers to find Mexicans. The light of interrogations. . . . My parents were nightclub owners in Mexico City. To me light is artificiality.[52]

More than a century after artificial light was used for the first time to provide a panoptic view, and Étienne Jules-Marey's polygraph was used for the first time by Italian criminologist Cesare Lombroso to assess a criminal suspect's blood pressure during an interrogation in 1890, artist Lozano-Hemmer creates aesthetic, collective, and connective experiences with heartbeat and lighting technologies.[53] Custom software, searchlights, sensors, scanners and other technologies play a major role in his biometric works. Here the ambivalence of technologies is explored; heartbeats are not simply a sign of life but signs with political connotations. And the ambiguities of surveillance are put in a balance that leans more toward care than control.

The use of artificial light for policing and military purposes has always fascinated Lozano-Hemmer, and he has articulated this with the deployment of surveillance technologies. Searchlights, a material often used in Lozano-Hemmer's pulse pieces, have historically been used in many contexts. They have illuminated the movements of people and military transport during war; they have blinded enemies with their brilliance; they have also prevented or controlled the transit of people, whether migrants and asylum seekers at borders or prisoners in jails. In addition to their military purposes to illuminate the enemy, searchlights have also been implemented for aesthetic and commercial purposes. [54] *Pulse Front: Relational Architecture 12* (2007), displayed over Toronto's Harbourfront, is one of Lozano-Hemmer's first pulse works, merging such surveillance and spectacular use of searchlights. The piece, like Abu Dhabi's later *Pulse Corniche* (2015; figure 6.9), is part of his *Relational Architecture* series, which he defined as "antimonuments for alien memory agency."[55] The artist proposed these public art projects to reconfigure public spaces and

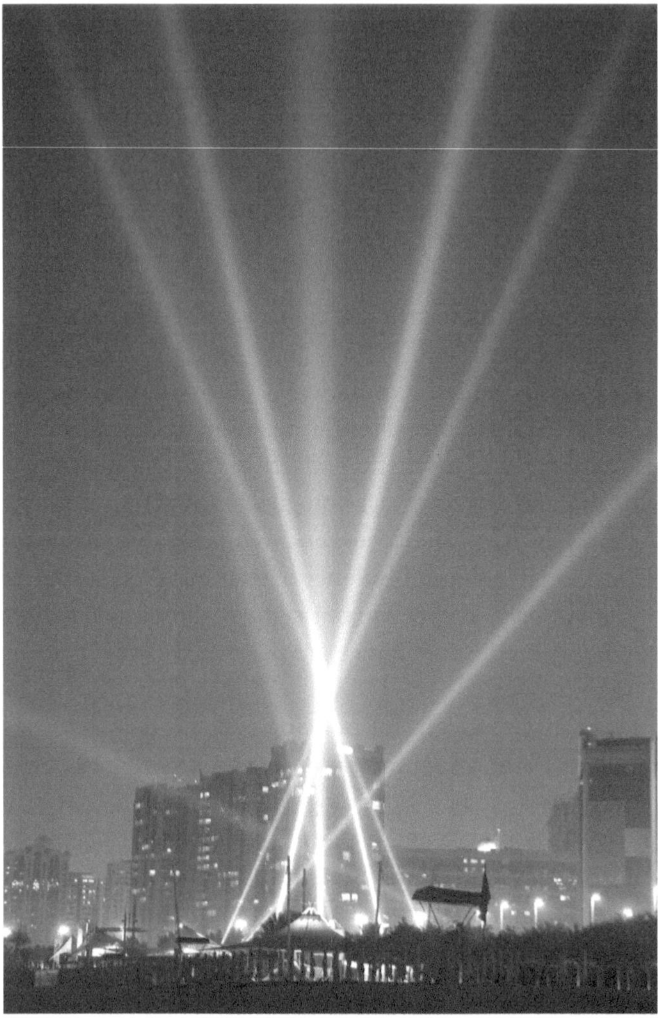

6.9

Rafael Lozano-Hemmer, *Pulse Corniche*, 2015. Guggenheim Abu Dhabi, United Arab Emirates. Courtesy of the artist.

create renewed forms of memories, relationships, and behaviors among those people who cohabit in a city.

Pulse Front and *Pulse Corniche* used a similar interface to *Pulse Room*, but instead of incandescent bulbs within a dark room, they used robotic searchlights projecting into the sky. The colossal interactive works use a matrix of robotic searchlights, a DMX controller, a computer, costume software, and metal structures with sensors located across different points. When passersby held the stick sensors, a computer read the data and sent it via USB cable to a minicomputer located in the base of the pulse stations, which translated the data into light beams via 10,000-watt searchlights. The lights traveled up to 10 kilometers across the sky. When the work was not being activated, the computer projected the biometric recordings of the last participants. The result was a spectacle of dancing lights in the dark sky that expressed life in a phantasmagorical reflection of pulsing beings.

A fundamental characteristic of Lozano-Hemmer's pieces is that they create a sense of community. The fostering of human connection via electronic art has been a continuing source of inspiration and investigation for his artistic practice. In his works based on the pulse, the artist prioritizes the creation of collective artistic experiences over solitary and individual encounters. This has become possible because of technological advances that have enabled the input of many people to be joined together and amalgamated into a single piece.

Lozano-Hemmer's most recent and most significant work using biometrics and searchlights aims to bring people together across the US-Mexico border. *Remote Pulse* (2019) is an interactive installation consisting of two heart-sensing stations, one located in Ciudad Juárez, Chihuahua, and the other in El Paso, Texas. Once they are touched, the sensors transmit the pulse of the stranger to the sensing station on the other side of the border, interconnecting spaces, persons, and hearts. Participants can feel the pulsing in their hands in a haptic connection. The piece was presented alongside *Border Tuner* (2019), which also interconnected the two sides through sensing stations that guided powerful searchlights. When the lights of two stations intersected, microphones and speakers automatically turned on and participants could engage each other in cross-border conversations.

Both pieces work in opposition to US usage not only of border searchlights but also of biometric technologies. These have been used to monitor and channel the intense back and forth of human flows across the border.[56] The exclusion of migrants from the political body indicates, according to Giorgio Agamben, a state of exception. This has become a norm found not only in the United States but around the world.[57] As noted by James C. Ross, the previously Orwellian state based on watching is now a cybernetic state control, a *bionetwork* that uses data recorders to decide who is part of the body politic and who it is necessary to exclude and "restrict their access to entry, benefits and jobs."[58] As part of these strategies, the US Special Forces designed Jetson and other heartbeat detector devices, which can be applied for both military and immigration control purposes. They can detect individuals by their cardiac signature from hundreds of meters away.[59] Many of these devices have been used to uncover migrants and asylum seekers hiding in trucks not only on the Mexico/US border but also on European borders. When aligned with technological forces, the heartbeat is immersed in socioethical quandaries.

Through heart reverberations, *Remote Pulse* creates a platform of dialogue for the questioning of political borders, the location of private and public frontiers, and the discriminatory categorization, objectification, and control of immigrants' bodies. The work explores the possibilities of expanding the body to unexpected and forbidden spaces via biometric technologies, which are often the domain of surveillance systems and are today a primary tool for distinguishing who is part of the body politic and who is not.

I think that *Remote Pulse* surpassed *Border Tuner* in its emotional capacity. People truly "felt" each other's pulses through the haptic interface, and this created an intimacy . . . it is as if you put your hands on the chest of another person to feel their heartbeat. I donated the piece to Ciudad Juárez and El Paso so that it can be permanently installed in a public space in each city and become a reliable connection . . . a bridge across Trump's wall.[60]

Regarding the use of technologies in the piece, Lozano-Hemmer commented in another interview,

I can't help but think about the ways in which these same biometric technologies are being used in ways I find appalling—for example, in the separation of families that's taking place at the border. For me, it's a beautiful opportunity to use these technologies of control and power to create connective experiences, poetic experiences, where we feel a sense of belonging instead of division and walls.[61]

In addition to fostering a space for spatial and poetic connection, *Remote Pulse* also fostered some empathetic and physiological connections. Recent neuroscientific studies demonstrate that the interoceptive sensitivity of heartbeats impacts both affective and cognitive empathy, and that people's cardiac rhythms sync when they connect spatially, whether they are sleeping, cuddling, or simply sensing their heartbeats for a certain period of time.[62] By facilitating the experience of touching the other from a distance and therefore experiencing a space of synchrony, closeness, and empathy, *Remote Pulse* emphasizes the interconnectedness and interdependence not just of persons but also of communities and border regions.

Lozano-Hemmer has created several works using heart rates that reveal interconnected technological systems and also use the heartbeat as a biometric identifier and as a vehicle for communicating the state of a person. During the COVID-19 pandemic, the artist created the online work *OnPulse* (2020).[63] Participants enabled their cameras so that their heartbeats could be detected using photoplethysmography. This optical technique detects blood volume variations by measuring light transmission. Once their heartbeats were translated into graphics and made audible as lub-dub sounds, users could access a chatroom. Here each heartbeat took the form of a flame, evoking the film *Macario*, and in that intimate and dark space, in that pulse room, one could initiate a dialogue with other hearts.

An earlier work also created for remote access was *Pulse Phone* (2009), an iPhone 4 app. It measured the user's heart rate using the phone's camera and flash, and could detect variations in the opacity of the person's finger via blood flows so that it could calculate their pulse in beats per minute. This information then appeared on the screen as an electrocardiogram. Lozano-Hemmer made the app by hacking Apple's camera code and developing an algorithm that analyzed the pulse. The app saved the pulse history so that the user could share it by email if they wished. The artist considered *Pulse Phone* to be a "trick device," with its only artistic merit being the aesthetic visualization of data as white lines on a red background. I suggest, however, that this app, and other self-track apps and gadgets such as the Apple Watch, Fitbit, Garmin, Oura, and Strava, show how pulse rate technological devices and apps have crossed the traditional boundaries between art, health, fitness, science, surveillance, and entertainment. The possibilities such devices offer are numerous. Some Strava users, for example, have located data breaches in the app and publicly revealed several US military bases and users' data and activities. Others have used Strava's GPS mapping to draw hearts, animals, and other figures in their running or cycling routes—an activity that is now known as "GPS art."[64] Using Fitbit body data, American duo LigoranoReese created *IAMI* (2013–2014), a series of portraits in which an individual's emotions and weekly Fitbit activities were displayed as a colorful and continuously changing fiber optic tapestry. The work considered how our lives, activities, and bodily functions are more than bits and bytes as they are threaded, measured, transmitted, and networked on a daily basis across the internet. And of course, our data is subject to commodification by companies or governments that seek the monetization of life.

Heartbeats combined with other biometrics have become a plausible secondary option by which electrical and computer engineers can verify identities in our hyperconnected world, even with their variability owing to mental states, physical activity, and age.[65] They have found the EKG signature to offer a secure system because it is a living password. This makes it more difficult to hack and more reliable when used in combination with other forms of authentication like fingerprints or facial recognition. As Donna Haraway anticipated in 1992 in her "Cyborg Manifesto": "We are all chimeras theorized and fabricated hybrids of machine and organism."[66] Surveillance technologies are today an assemblage of technological and organic inscription devices, intersecting various parts and processes that are disaggregated for the purposes of observation. Data from computing devices, mobile phones, CCTV cameras, facial recognition, tagging systems in social media, and mobile applications that capture heart rates can be tools of control with which governments can expand their territories of dominance.

Today, the transformation of the body into data is ubiquitous. This information has been particularly useful in medicine and law enforcement, two arenas that are increasingly interconnected. As N. Katherine Hayles has stated, information has "lost its body," as it is dispersed into a diverse technological system that defines who we are and how we are perceived.[67] Technology has become so "entwined with the production of identity that it can no longer meaningfully be separated from the human subject."[68] The subject has become "a collection of heterogenous components, a material-informational entity" in constant flux.[69] According to biometric researcher Irma van der Ploeg, graphs and abstract images of the body have changed the ontology of the physical body into "flows of information and communication patterns" captured and stored by IT technologies. These databases of information have the potential to attach a "biometric identifier to every personal file in these databases" for use in the control of citizens, particularly in the areas of health, migration, criminal investigation, insurance classifications, and intelligence services.[70] Furthermore, as observed by Kate Crawford, as AI is designed to "benefit the states, institutes, and corporations that they serve,"[71] it is a form of power that has resulted in more algorithmic and predictive technologies that amplify existing structural inequalities. As a result, our bodies, and thus heartbeats, have become data that algorithmic systems might use and misinterpret.

In Lozano-Hemmer's *Pulse Index* (2010; figure 6.10), participants are framed not as body entities but as abstracts: as graphic recordings of heart rates and fingerprints, extracting the body from its physicality into an amalgam of virtual data. The bodily landscape of *Pulse Index* is activated when visitors touch a custom-made sensor equipped with a digital microscope. An image of the fingerprint is taken from the microscope itself and immediately projected, pulsing at their own heart rate, with an image of their cardiac waveform. The work can record and display the data of up to a thousand participants, but as more people activate the piece, the recorded images gradually become smaller and smaller until they vanish. This immersive installation highlights the human dimension through technological paraphernalia but dispenses with the notion of the individual. The pieces are most effective when the individual input disappears and merges into the community.

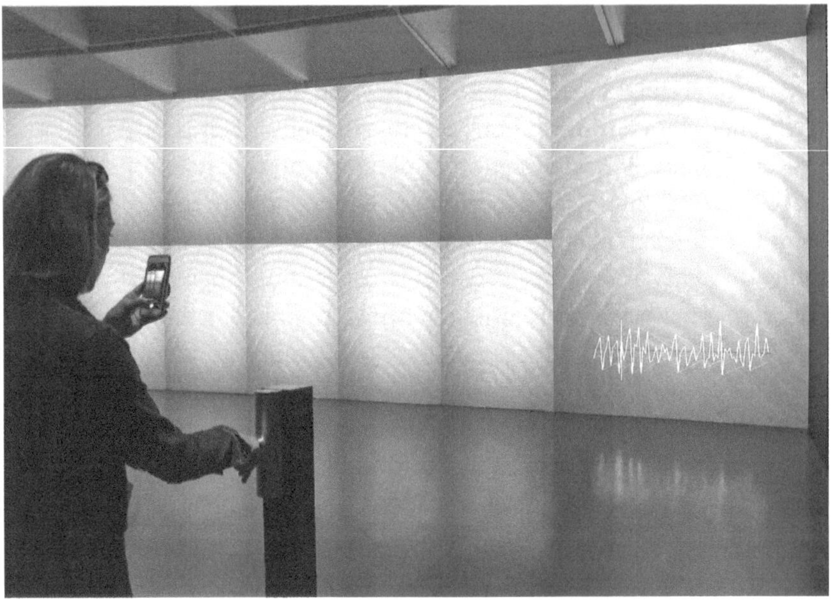

In its beguiling landscape of palpating traces, heart rates, and fingerprints, *Pulse Index* oscillates between a life form and a web of information, as an organic entity coming into contact with a nonorganic one. The life form is created through pulsating graphs and fingerprints and by a pulse that gives life to an image usually collected just for official identification and surveillance purposes. The web of information alerts us to the mechanisms of control, which we are already embedded within, and the uses of this body of information for social categorization and discrimination. But it also reminds us that we are part of a collectivity that lives in an

6.10

Rafael Lozano-Hemmer, *Pulse Index*, 2010. Hirshhorn Museum and Sculpture Garden, Washington, DC, United States, 2018. Photo by Cathy Carver. Courtesy of the artist.

inescapable hybrid system, where a new form of body has been created and governed by the conjunction of both body and data.

While biometrics, computers, networks, programs, and other technologies can be used to intrude into our privacy, artists like Lozano-Hemmer subvert their ontology by using them as creative tools and transforming them into a valuable resource that promotes community and aesthetic experiences. Heartbeats and lights in Lozano-Hemmer's works can be read as paradoxical material: they join the biological with the technological, as waves that represent absence and presence, life and reason, and as materials, they are used for both care and control, intimacy and spectacle. In the pulse works, lights and pulse commingle in a reflection of our nonbinary existence, showing that things are not always one thing or the other but can be many things at the same time. But most importantly, they are intangible elements that produce the visual illusion of the extension of the energy of our bodies into unimagined spaces; this is done by creating connective and collective experiences that extend the vitality of our intimate hearts into the public sphere, reminding us of their multiple connotations and (mis)uses.

7

REGENERATIONS

The heart is a unique, powerful, and resilient organ that is both courageous and sensitive. Despite trials, illnesses, misuses, and tribulations, it keeps beating, strong, loyal, and persistent, clinging to life. The heart has extraordinary capacities of adaptation and regeneration through its ability to replace damaged or lost tissues and through its plasticity. This means the heart can change size and adapt in a relatively short period of time; for example, it can double its size during pregnancy, and an athlete's heart can become thicker than the average heart.[1] Heart regeneration also refers to a person's efforts at emotional or spiritual healing. We may feel shattered or damaged after experiencing heartbreaks or emotional stress, and we seek ways to heal and move on, whether through counseling, self-care, art creation, or other means.

In this final chapter, I engage with artists who have resorted to listening to their heartbeats to heal from both physical and emotional wounds. In her art and life, Yoko Ono conceives the steady beating of our hearts as an important source of regeneration and as a connective symbol. This same idea runs through the work of Linda Montano and Christian Boltanski, artists who are both addressed in this chapter. While Montano listens to her heartbeat to overcome personal crises and breakups, Boltanski uses heartbeats to remember the victims of war and the lives that will be gone, leaving no trace, no memory. He confronts us with the ongoing regeneration of life and death through an archive of heartbeats.

Historically, the brain has been gendered as masculine and the heart as feminine. "Think with one's heart," and thus listening to the heart prior to the brain, has been used to belittle the way women think. But in the works of Yoko Ono, Linda Montano, and Christian Boltanski, the act of listening attentively to our heart thumps indicates the unique and undoubtful potentiality of heart rhythms as language, presence, guide, knowledge, and remediation source. The heartbeat revitalizes, nourishes, and purifies the entire body; just as the circulatory system connects with each corporeal cell, the heartbeat ties us to ourselves and, by extension, to others.

GETTING CLOSER: YOKO ONO

While visiting a gallery with a friend, you notice a short-typewritten invitation that reads, "Listen to each other's pulse."[2] As you consider the instruction, you and your friend are briefly united by what Yoko Ono celebrates as one of the ultimate signs of connection and introspection.[3] *Beat Piece* (1963) and *Pulse Piece* (1963) are two of the first of her heartbeats works. Both are event scores—namely, invitations to execute or imagine whimsical actions, which "provoke a mental image that defies banal sense and so challenges one's imagination to see the world from a radically different perspective."[4]

PULSE PIECE
Listen to each other's pulse by putting your ear on the other's stomach
1963 winter

BEAT PIECE
Listen to a heart beat
1963 autumn

Yoko Ono has constantly claimed that, for her, "art is a means of survival."[5] She has overcome the aftermath of World War II, a suicide, a voluntary stay in a psychiatric hospital, the loss of a child, John Lennon's murder, and the sustained racism and misogyny of the public. There is no more compelling sign of the continuation of a life than the beating of the heart. Heartbeats are part of a person's quotidian experience; they are our faithful companions during our brief existence. It is no surprise, then, that cardiac rhythms have offered a shared language that Ono has consistently used to try to reconnect with herself and others to make a better world.

Deeply thoughtful yet simple, Ono's heartbeat works have taken several different forms. Some are silent music pieces in the form of short poetic texts that encourage people to engage in auscultation or the listening of body sounds; others are audible experimental music created at decisive moments of her life; a few are sculptures that, through a careful selection of objects, relate heartbeats to the passage of time. Although varied, many of Ono's heartbeat works are derived from the same nucleus: her book *Grapefruit.*

Grapefruit (1964) is an enthralling and influential artist book that collects Ono's early instruction works, and which can be considered the heart of her art.[6] The book, which served her as therapy at a time when she was experiencing mental health problems, is an imagination and care exercise.[7] Through drawings and textual forms evocative of *haiku*, the artist gives simple yet unusual and playful instructions, all related to daily life, nature, and interpersonal relationships. Their execution is always left up to the reader's imagination. *Blood Piece* (1960), for instance, invites the reader to use blood as paint to allude to art and life, and the absurdity of taking things too seriously.

> **BLOOD PIECE**
> Use your blood to paint.
> Keep painting until you faint. (a)
> Keep painting until you die. (b)
> 1960 spring[8]

Grapefruit—a fruit that is a "hybrid of lemon and an orange"[9]—reflects Ono's feelings of being a "spiritual hybrid,"[10] always living between two countries from an early age. Born in Tokyo in 1933, Ono spent most of her early life studying and working between Japan and the United States. In 1941, her family sailed back to Japan after the attack on Pearl Harbor, where Ono used her vivid imagination to escape from the horrors of war.[11] Since then, imagination has become a core part of her life, work, and music.

Ono's heart was set on music since her childhood, as her aristocratic education included musical training, studying the piano, harmony, and composition from an early age.[12] Her father later dissuaded her from pursuing a career as a composer because "the field was too hard for women."[13] As an alternative, Ono studied in the philosophy department of Gakushuin University before moving to New York in 1953, where she studied music composition and literature at Sarah Lawrence College from 1953 to 1956. During this time, years before her encounter with American experimentalism, Ono actively created haiku, poems, novellas, and scores to fight the feeling of isolation in her new hometown.

In 1956, Ono became acquainted with the artists who would become the pulse of the Fluxus movement. She herself became a formative figure in Fluxus, not only as an artist but also as an organizer of various experimental events in her Chambers Street loft. Ono was part of a wave of artists who, during the 1950s and 1960s, began to interrelate art and life, and to incorporate ostensibly invisible and inaudible energies into their music, words, and actions. Like many of her contemporaries—such as John Cage, who experienced the sound of his own circulation in an anechoic chamber—Ono's aesthetic and musical sensibility explored silence and the internal sounds of the body.

At that time, the postwar generation was committed to alternative ways of approaching life away from Western modes of rational, controlling, and pragmatic thought. They wanted to move away from abstract expressionism and the commodification and institutionalization of art. The avant-garde artistic scene became interested in Zen, particularly in Manhattan, where many Fluxus and Happening artists attended D. T. Suzuki's lectures on Buddhism at Columbia University. These events were a sensation with a generation that was dedicated to the spontaneous and the simple instead of grandiose and worldly goals.[14] At the invitation of John Cage—an artist who encouraged Ono to continue her experimental work—she attended Suzuki's lectures, but she had been exposed to Zen from an early age. As the art historian and curator Alexandra Munroe has observed, besides the already recognized influence of John Cage in spreading Zen ideas in the arts, "Ono's experiential and intuitive art played a key role in the transmission of Eastern aesthetics to an international avant-garde trained in post-war existential thought."[15]

Beat Piece and *Pulse Piece* are compositions that combine Western and Eastern ideas. Listening to a person's heartbeat has been a long-standing practice in getting to understand their health and emotional state. As explained in chapter 1, since ancient Chinese and Greek medicine, physicians touched the skin to reveal the body's inner secrets.[16] According to Shigehisa Kuriyama, the art of measuring the pulse with the haptic method transformed the sounds of the body into a language.[17] Ono's pieces echo this idea of heart rhythms as a hybrid of sound and language, demonstrating the power of auscultation to connect not only to an individual but also to a society.

Conceived as music but also as an event, *Beat Piece* was first performed at *Perpetual Fluxfest*, an evening organized by Fluxus central figure and coordinator George Maciunas, held in 1965 at the Cinematheque in the East End Theater in New York. It was performed by Nam June Paik, Shigeko Kubota, Tony Cox, and Bill and Mimi Waring, among others.[18] On this occasion, participants lay on top of each other's bodies to hear their hearts beating. Similarly, *Pulse Piece* invited the audience to hear one another's pulse via the stomach through the abdominal aorta. The simplicity of Ono's instructions underlines that she gives more relevance to the effect of interactions than to the object or the command itself. "Form is emptiness, emptiness is form," claims the Heart Sutra, one of the most popular Buddhist scriptures. It refers to the interdependence of all things, particularly of the physical and the mental. Listening to the heartbeat makes a link between the body and the mind since the heart is in constant relationship with the mind. Ono's musical compositions are effectively silent. They intend to induce sound in the imagination and require concentration in order to be appreciated: "I think of my music more as a practice (*gyo*) than a music. The only sound that exists to me is the sound of the mind. My works are only to induce music in the mind of people."[19] The value of silence comes from Ono's knowledge of Zen philosophy, which considers silence and emptiness as fundamental values in everyday life, as fields of possibilities, and as forms of nonverbal communication that convey knowledge, emotions, and truthfulness, and help to produce a consciousness of oneself and of the world.[20]

The invitation to stop all thoughts and activities and to focus on the heart beating—either as a real pulse sound in the chest or an imagined sound that is heard through one's inner perception—is a call to appreciate life itself and to restore ties between people. Ono's works have a meditative character and are created by following everyday actions in a contemplative and imaginative way. At the core of her works is imagination and the power to create possibilities and to apprehend perspectives and feelings different from one's own.

Ono's conception of cardiac sounds as music is part of a long thread going back in history. Chinese and Kampo medicine used poetic and imaginative forms for the pulses that were felt to be straight, slow, long, taut like a musical instrument string.[21] Just like Yoko Ono, Chinese doctors feeling the *mo* used poetic expressions to describe it: "like clouds floating in the sky" or "quiet and whispering like falling elm pods."[22] Galen and Boethius, in the early sixth century, established a correspondence between the heartbeat and musical ratios, and Laennec, the father of indirect auscultation, imagined he heard orchestras while he was listening to patients' hearts. For centuries, physicians and philosophers have used the imagination to associate heartbeats with musical sounds and compositions, and Ono's musical instructions continue this tradition of assigning musical connotations to the pulse. By inviting people to listen to the drumming of the heart in different parts of the body, Ono induces a space of consciousness of the mind and the body, and highlights the interconnectedness of the different parts of the body. But more importantly, by instructing people to perform a direct auscultation—a practice which has been replaced with instruments that avoid physical contact—she invites people to break with convention and get closer.

A similar concern appears in another of the artist's works also created in 1963. For *Touch Poem for Group of People* (1963), Ono invited people to come together and touch each other. This was an invitation to cross the conventions of Japanese and Western cultures that share a culture of distance and no touching. In an interview, the artist explained,

> All my stuff was not really to do with body. Touch could've been very conceptual too. But also, at the time, I really wanted people to touch each other physical because we never did that, you know, in those days, or even now. It's a pity we're getting not closer, but more distant to each other.[23]

Closeness, that quality of being at a short distance, not only spatially but also intimately and affectively, is at the core of Ono's pieces. All her heartbeat instructions are a call to approximate, regenerate, and appreciate life, and this can be achieved through the subtle but magical act of placing your ear on someone's chest or stomach to listen to their inner rhythm. Touching is not only about putting one's own skin in contact with another person's skin; touch implies understanding and being affected. As both Derrida and Nancy argued, the meaning of touch "must touch on the untouchable."[24] The liminality of touch for Derrida lies in the interactions between objects that are either possible or impossible. Although, as Nancy perceives, touch is ultimately always touchless, it is "the proximity of the distant, the approximation of the intimate."[25] So, touch is more than the mere contact between things; it has an ontological quality that marks beings by the being-with. Listening to someone's heart, as in Yoko Ono's instructions, is to make contact while remaining apart; it is to touch the heart without touching it. And it is in itself an act that links to the untouchable because as Jean-Luc Nancy states, an "experience at a heart: it is everywhere, it is nowhere."[26] Thus, the leaping of the heart unites the inside and the outside, the soul and the body, nature and culture; they are an experience that speaks of us and through us. Touching the pulse, then, is not only an act of knowing the physical condition of the body; it is also an action that connects us with others. When Ono's haptic instructions are followed, they create a communication channel that opens the possibility of sensing and reflecting on communality, togetherness, and closeness.

The simplicity but profundity of Ono's abstract and aporetic instructions is reflected in the artist's interest in event scores. By the distillation of a musical score into short, enigmatic texts, similar to instructions, they indicate the actions to be performed.[27] Here we can also see connections with Dada poetry and its embrace of the "everything and the nothing," as well as with Eastern Zen *kōans*, *nō*, and *haiku*. Haiku are short and intuitively made poems that invite readers to question their assumptions about reality. They are like all of Ono's instructions: seemingly "unfinished" poems that require the reader to complete them by experiencing them in their mind. "Listen to a heart beat" is not a seventeen-syllable haiku, as they usually are, but a six-syllable haiku that is pleasing to both the ear and the mind. It is a short sentence that expands the imagination, which, like the heart, has flow and rhythm, and evokes images of the natural world.

All heartbeat haiku by Yoko Ono fit well in the Fluxus aesthetic because they involve improvisation, simplicity, and the everyday, all represented in their most essential form: the pulse. Ono's sensitive practice is focused not on the artwork itself, however, but on the connection to the self through common actions. On one occasion, trying to differentiate her work from other Fluxus artists, like Allan Kaprow, whose art simulated life, Yoko Ono declared, "Art is not merely a duplication of life. To assimilate art in life is different from art duplicating life."[28]

Ono appeals to a connection of mind and body, self and world, and art and life through various forms of contemplation. For her, the everyday is an experience that can be a work of art. This idea is reflected in one of her first sculptural pieces, in which she merges time and heartbeats. *Eternal Time* (1965; figure 7.1) was first shown publicly in 1966 at her *Unfinished Paintings and Objects* show at London's Indica Gallery, in which she translated some of her instruction art into objects. The sculpture consists of a small Plexiglas box containing a gold-colored clock with a stethoscope attached to the outside. The watch, which has the title of the work engraved on it, only has a hand for counting seconds.

Continuing a long tradition of relating heartbeats to units of time, *Eternal Time* is a materialization of *Grapefruit's Clock Pieces*, in which she invites us to listen to time by creating an imaginary system of counting. Here Ono requests that we listen carefully to the pulsating and timeless rhythm of the universe. For Yoko Ono, there is no need "to stick to the two-dimensional or three-dimensional world. In your mind, you can be in touch with a six-dimensional world if you wish."[29] So in the ticking of a clock heard through a stethoscope, mind and body, circadian time and cosmological time can intermingle in an imaginative act. Therefore, the art of this piece lies not in the object itself but in the act of listening to the heart as the passage of time.

The exhibition at Indica Gallery was very significant for the artist, not only because it was one of her more complex shows but also because during the preview of the show, she met John Lennon for the first time. Ono

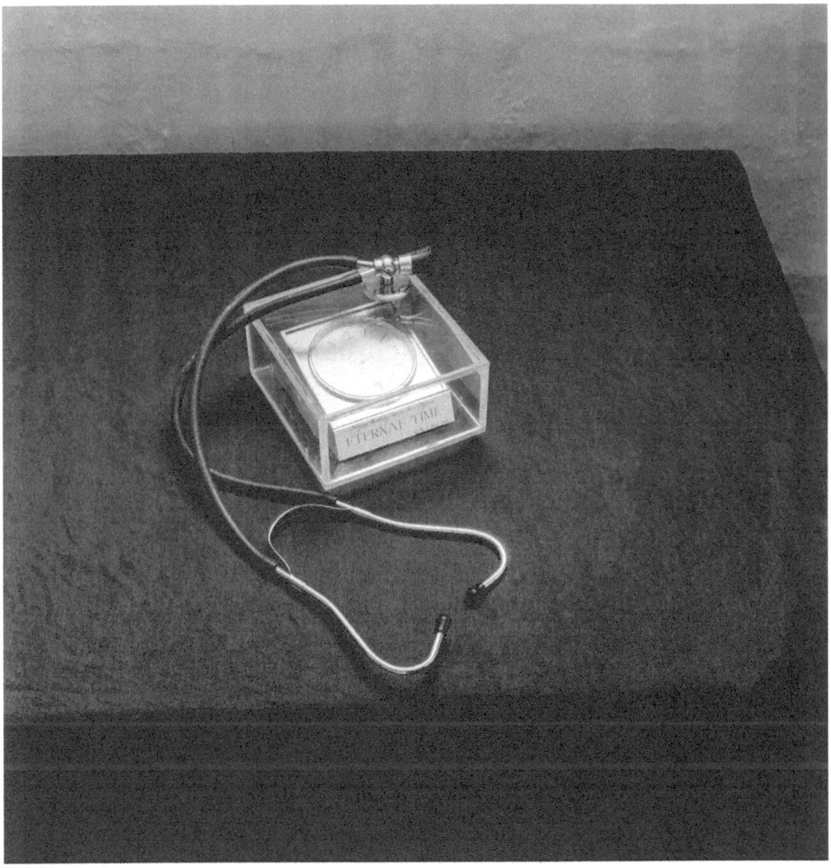

7.1

Yoko Ono, *Eternal Time*, 1965.
Gilded clock with second hand only, stethoscope, clock, Plexiglas bonnet.
© 1965 Yoko Ono. Courtesy of the artist.

and Lennon married in 1969, becoming one of the most famous and most criticized couples on the planet. During their lives together, they influenced each other's work and pushed each other's creative boundaries. The couple's more successful collaborations were three intense albums: *Unfinished Music No. 1: Two Virgins* (1968), *Unfinished Music No. 2: Life with the Lions* (1969; figure 7.2), and *Wedding Album* (1969). The three records, some of which contain heartbeats, merged Lennon's interest in pop and rock music with Ono's conceptual practice.

The couple's collaborations were rooted in the art and life aesthetic: their albums were a sharing of their intimate life, which included not only revelations of their life together but also their most intimate vital signs. On the initiative of Ono, who always included heart rhythms in her works, two of these records engage with heart sounds. It was also Ono's idea to name two of the albums "unfinished music," which suggested, as with all of her instruction works, "a participatory music." Listeners were invited to complete the songs in their own ways, such as adding heartbeats and emotions or imagining extra tracks.[30]

On the front cover of *Unfinished Music No. 2: Life with the Lions*, Ono appears sad, lying in a hospital bed with John sitting on the floor by her side. The intimate album was mostly recorded in a room at Queen Charlotte's Hospital in London, where Ono suffered a miscarriage on November 21, 1968. "Baby's Heartbeat" records the heart palpitations of the unborn baby in Ono's womb. The recording came from a phonocardiogram that the doctor took at around the third month of her pregnancy. Using heartbeats from the womb of a mother in music was something new; it wasn't even possible until electronic fetal phonocardiography became commercially available in the medical sector and allowed some parents, like Ono and Lennon in the 1960s, to have a record of their baby's heartbeats. As noted by Anne Pollock, "In the sonic effect of the foetal heartbeat, there is an effective and affective articulation of an individualized foetal heart."[31] This is then conceived as the "phantasmic origin of personhood."[32] In Ono's and Lennon's album, the three minutes of the infant cardiac rhythms are followed by a two-minute silence honoring the short existence of the couple's first son. "Baby's Heartbeat" is a pioneer work in the use of heartbeats, a song of mourning, a memory, which turned a short life into timeless art as an assimilation of art in life.

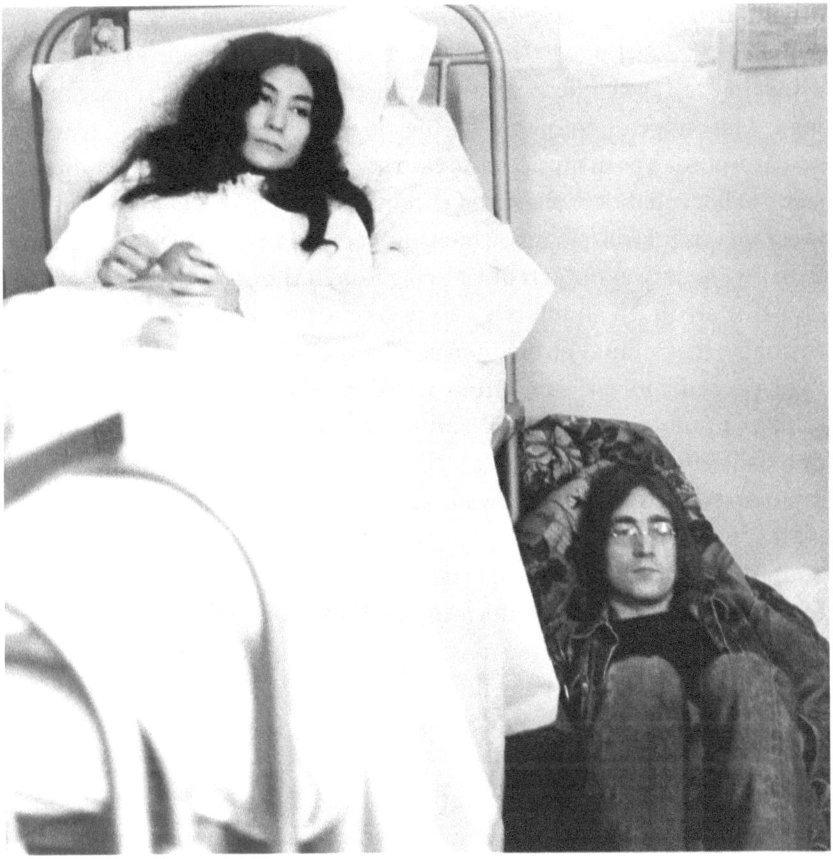

7.2

Front cover of John Lennon and Yoko Ono, *Unfinished Music No. 2: Life with the Lions*, 1969. Yoko Ono/John Lennon. Zapple Records. Photo by Susan Wood. © Yoko Ono. Courtesy of the artist.

Another moving track employing heartbeats is "John & Yoko," which is part of the *Wedding Album*. The song features Ono and Lennon's palpitations while they call each other's names at various tempos and volumes, expressing a wide range of emotions, such as joy, tenderness, anguish, fear, and sexual arousal. For many decades, Ono's work was not properly credited in some of her collaborations with Lennon. "Imagine," one of John Lennon's most successful solo singles, which has been an anthem of peace for more than fifty years, was only credited to Lennon until 2017, when Ono received a writing credit. In an interview in 1980 on BBC radio, Lennon declared, "Actually, that should be credited as a Lennon-Ono song because a lot of the lyric and concept come from Yoko. But those days I was a bit more selfish, bit more macho, and I sort of omitted to mention her contribution." A useful example of her contribution is to be found in "John & Yoko," a sonic version of Ono's *Body Sound Tape Piece* (1964),[33] which instructs performers to make recordings of people "crying, longing, excited, calm, doubtful, etc."[34] Yoko Ono once declared, "All my things have to do with the inner life, inner communication and transformation."[35] This song is no exception, presenting the range of feelings that a couple share and the beats that always accompany them.

In their albums, Ono and Lennon introduced a musical style that was neither pop nor conceptual but experimental sound art. While Lennon discovered conceptual experimentation through Ono, she in return found in him a helping hand to realize her ideas of inaudible music in sounding music. This is why in these records, heartbeats are not a silent element as in her instructions, but a real audible sound of life.

For Ono, drums, including Japanese *taiko*, are "the heartbeat of the universe. The drummers are letting us touch the heartbeat of the universe through their drumming."[36] In her albums, she clearly perceived a connection between the pounding, hefty rhythm of rock music and heartbeats.[37] In an interview published in the 1970s, she considered the bass and drums to be a response to the lost heartbeat in music:

> I realized that the classics, when they went from 4/4 to 4/3, lost the heart beat. . . . It's as if they left the ground and lived on the 40th floor. Schoenberg and Webern—Webern's on the top of the Empire State Building. But that's all right. Our conceptual rhythm got complex, but we still have the body and the

beat. Conceptual rhythm I carry on with my voice, which has a very complicated rhythm even in "Why," but the bass and the drum is the heartbeat. So the body and the conceptual rhythms go together. These days I'm putting a beat under everything I do.[38]

Heartbeats have continued to appear in Ono's most recent works. For *Heartbeats/Syria* (2012), she permeated the Serpentine Gallery with a constant beating sound, which visitors might at first associate with notions of birth and vitality; however, once they encountered the piece's title, it brought to mind the millions of people suffering through the ravages of war. As someone who experienced the horrors of warfare up close, Yoko Ono has dedicated part of her art and life to anti-war activism. Take as an example the billboard campaigns *War Is Over (If You Want It)* and *Bed-In for Peace* (both 1969) cocreated with John Lennon. Here the couple spent a week in bed wearing white pajamas as a form of pacifistic protest. *Heartbeats/Syria*, like her previous anti-war work, is a call for empathy. As she has repetitively affirmed, "Our hearts beat in unison. Even when we fight with one another. We are indeed one world, one people."[39] In these pieces, the heartbeat is political, as it relates to emotions and feelings that underpin notions of community. Heartbeats bind individuals, help them to sense themselves and others, even to grasp the relentless rhythm of the universe. In Ono's words, "Listen to the heartbeat of all things on the planet. It is symphonic."[40]

Once described by John Lennon as the world's most famous unknown artist, Yoko Ono is a pioneer of conceptual art and of the use of heartbeats as media in the arts. In the face of trauma, adversity, and illness, rather than opting for anger or sadness, Ono has turned to the heartbeat as a form of bonding, remediation, and love. Her secret to emotional health and the source of her admirable strength lies, she says, in "something that makes your heart dance. . . . It's that little dance of my heart that I listen to."[41]

HEART ENDURANCE: LINDA MONTANO

A woman dressed in shorts and equipped with a stethoscope is alone inside a gallery. She is American artist Linda Montano, who spent three days locked in the Camerawork Gallery in California. There, she taped a stethoscope to her heart and wore it during her whole stay to literally and symbolically listen to her *Heart Murmur* (1975; figure 7.3).[42] A heart murmur is an unusual whooshing noise produced by the closing of the valves and a turbulent blood flow. It is heard between heartbeats and alters the normal lub-dub sound of the heart. Cardiologists distinguish between two types of murmurs: innocent murmurs, which are mostly considered harmless, but still imply the possibility of future heart issues, and abnormal murmurs, which can indicate serious underlying heart conditions. Believing that "the purpose of art is to create a silent mind and open heart," Montano immersed herself in the silent gallery to listen carefully to the congenital heart murmur that has accompanied her throughout her life and perhaps find inner guidance at a transitional time.[43] Montano described the motivation for this in an email:

> Because I always make art about my life, I found a way to use my heart as a vehicle for alchemical transformation and as a pointer to the need to gather information from this center of emotional truth. At the time I was in a very deep state of inner turmoil, leaving a marriage and entering a new relationship. May hearts help.[44]

Born in New York in 1942, Montano was raised in a Catholic household by her parents, who were artists and orchestral musicians. At the age of nineteen, after suffering a sexual trauma, she joined the Maryknoll Sisters order. Two years later, after presenting with a severe case of anorexia, a common "comorbidity" in sexually abused patients, she left the convent and began to study art.[45] She first graduated from the College of New Rochelle and later earned master's degrees from Villa Schifanoia in Italy and the University of Wisconsin. However, her heart always pursued monastic life: Montano lived for a number of years in the Ananda Ashram where she studied yoga and meditation, and for two years at the Zen Mountain Monastery in New York with her then-partner Pauline Oliveros.

Linda Montano's artistic practice—which is permeated by rituals, discipline, seclusion, and community caring—is inspired by her interests in Zen meditation, yogic philosophy, Catholicism, and her time as a novice.

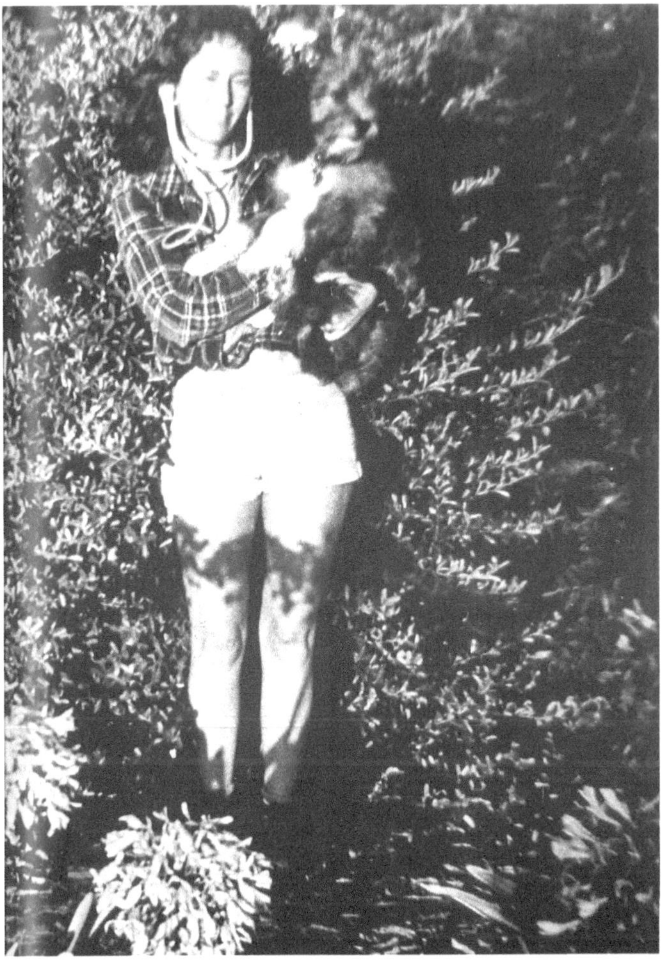

7.3

Linda Montano, *Listening to My Heart . . . A Congenital Murmur [Heart Murmur]*, 1975. Photo by Mitchell Payne. Courtesy of the artist.

She started exploring the observance of her daily life, the conception of art as life, and the openness to feel and hear her cardiac thumps in the 1970s. At this time, she was influenced by the work of artists such as Allan Kaprow, John Cage, Pauline Oliveros, and Barbara T. Smith, all of whom were inspired by Zen philosophies and experiences of life that were aesthetically and symbolically charged.[46]

Montano's work is considered part of a group of artists from the 1970s and 1980s who are often referred to as endurance artists. They explored art and life and challenged the human body by testing its physical and psychic limits through endurance and durational actions.[47] Some of the strategies of endurance artists included isolation for long periods of time, the infliction of pain, or exposing aspects of their intimate lives. As historian Karen Gonzalez Rice notes, "In the face of physical and psychic extremity, the endurance artist simultaneously embodies ethos and pathos, death and survival, vulnerability and discipline, victimhood and heroic agency."[48] It is no surprise, then, that many endurance and durational artists have used heartbeats as artistic material. Heartbeats represent a direct response to physical and psychological factors, to death and survival, as well as being a testament to time passing, to being alive, and to the risks of being a social living creature.

One of the most extreme works in which heartbeats and endurance are at stake is Jack Goldstein's *Untitled (Burial Piece)* (1972). For this performance, he had himself entombed at night on the top of a hill while wearing breathing tubes and a stethoscope that sent his heartbeat's signal to a blinking red beacon above ground. In addition to questioning systems of representation and the self through this symbolic death, the artist shows the misery of being dead while alive. Marina Abramović and Ulay also played with risk, survival, and emotions in *Rest Energy* (1980). Over a period of four minutes, Abramović held the hand grip of a hunting bow while Ulay held the string with an arrow that pointed directly at Abramović's heart. During the performance, both gradually leaned away, increasing the tension on the bowstring. Small microphones attached to both of their hearts recorded their pulsations, which, as the seconds passed, became more intense, mirroring the relation of mental and emotional strain and physical responses.[49]

As an artist who has been interested in inner life, secrets, and taboos, Montano uses her heartbeat differently than many of her endurance colleagues mentioned above. She does not envisage the heartbeat as a form of representation, nor as a physiological sign that only responds to emotions, but as the fodder of her art practice and as a therapeutic hope. *Heart Murmur* is one of several emotional autobiographical performances that Montano has created in her search for personal and spiritual transformation. It was conceived as a time-based action that required endurance, patience, and self-knowledge. The work consisted of living in the gallery under the gaze of visitors.

The only documentation available of the performance is one black-and-white photograph. Here Montano is in a garden holding her dog Betty while listening to her heart with a stethoscope. The image is published in her book *You Too Are a Performance Artist*, in which the artist mentions that sound artist Pauline Oliveros attended the opening of the performance disguised as a suburban woman. *Heart Murmur* was produced at a difficult time during which Montano was conflicted about ending her relationship with her ex-husband, the artist Mitchell Payne, and starting a new relationship with Oliveros. "I was in an incredibly compromised situation: leave him? Go to her? I had no idea what to do," she recalls.[50]

And as noted by Donna Haraway, "Having a dog lowers one's blood pressure and ups one's chances of surviving childhood, surgery, and divorce."[51] Betty was Montano's heart-love and her constant companion during her breakup and in a performance from which she said she only learned one thing: "that the heart is hungry and murmurs."[52]

Heart murmurs have been a constant in Montano's life, not only as a personal physical anomaly but also as a symbolic pain, a trauma. This is the territory that her art has helped her to navigate and understand. Montano considers that three events in her life caused the emotional abnormalities in her heart:

> When I was about 10 my grandmother said to my mother: "there is something wrong with Linda here," as she touched my left chest, near my heart. When I was first abused sexually, the perpetrator groped my right breast . . . my heart. Those two events gave me a symbolic murmur.[53]

Another acute pain was the unexpected death of her ex-husband, Mitchell Payne. Their breakup had inspired *Heart Murmur* several years before his tragic death from a gun accident in 1978:

> When he was murdered, my heart murmured very loudly. I had divorced him because my heart could not love the only person who ever really loved me. My heart was too wounded by my personal story. This tragedy has cemented me to the heart of everyone's tragedy. Now I am not an empath but a seeker of compassion.[54]

By emphasizing a movement from empathy to compassion, the artist positions empathy (drawing from the Greek *em*—"in" and *pathos*—"feeling") as something passive. It implies only the ability to feel and observe another person's emotions, while compassion implies the ability to be mindful of suffering and a desire to help, to find a cure for pain. Montano has always tried to establish a dialogue with her heart, to reach her most inner self and find emotional health. No matter how broken her heart has been, it still keeps beating, proving to be the greatest source of her resilience.

In her durational performances, Montano has always used her body and her private life as material, often drawing on different forms of tethering and confinement. For the three-day performance *Handcuff* (1973), Montano tethered herself to artist Tom Marioni; for *Living with Pauline Oliveros in the Desert for Ten Days* (1975), the two artists stayed together continuously for the set period, considering every single thing they did as art; and for her very challenging *One Year Performance* (1983–1984) with Tehching Hsieh, they shared every aspect of daily life from going to the supermarket or catching a bus to watching movies.

In *Heart Murmur*, Montano uses the stethoscope as a form of self-tethering that connects from a distance and forces her to listen to herself. Like its inventor René Laennec, Montano used this medical instrument to separate her mind from the outside world in order to perceive her interior through her heartbeats. But while people in the nineteenth century feared that the device could reveal the heart's most inner secrets, Montano uses it with precisely those magical and telepathic overtones in mind. She listened to the murmur of her heart as an act of sympathetic magic—that is, as an action that aimed to communicate with her heart and to receive advice from her own organ as a means to healing.

For millennia, emotions, hearts, the self, and medicine have been connected. In fact, the image of the heart as a symbol of affect originates in the ancient Greek medical tradition associated with Galen. This was based on humoral balances and for centuries asserted that the origin of maladies was connected to emotions until William Harvey discovered the mechanisms of blood circulation. These later led to the development of technologies that revealed more clearly how the organ worked. Closer to our own era, the rise of cardiology, psychology, and now neuroscience have positioned the heart as a passive receptacle of the brain's desires, although since emotions influence our cardiac activity, and we feel this in our heartbeat daily, the heart continues to be deemed an organ by which we "feel" and "know." It serves as a repository of the self and the ultimate source to connect and know about ourselves. This is why Montano connects through her heart to reach her feelings and so gives more prominence to emotional truth than to rational reasoning

In the 500s, Saint Benedict wrote, "Listen carefully . . . and incline the ear of your heart."[55] With those words, the founder of Western monasticism aimed to pay more attention to the emotions lodged in our vital organ than to words and actions. The relationship between the heart and ears in Montano's work is not only symbolic but physical. Today, we know that the ear is a window to heart health, as cardiovascular diseases can affect a person's hearing owing to inadequate blood flow. When people are heartbroken by the loss of someone they love, they can present with what doctors call a stress-induced cardiomyopathy (also called "broken-heart syndrome"), which causes low blood pressure.[56] Furthermore, hearts have even been conceived as harboring a pair of ears: physicians in the fifteenth century named the atriums of the heart *auricles* because of their resemblance to dog's ears.

In Montano's *Heart Murmur*, as noted by Charles Stankievech, "the space between the ears is pounded out to create a space for another interior space: the heart. One organ is remapped onto another organ, one instrument onto another, one space into another space."[57] And it is in this neutral space where Montano aims to settle her inner dispute in a moment of turmoil.

Montano's search for communication between an inner and outer condition of the body and the psyche through the vital organ can be connected with Frida Kahlo's *The Two Fridas* (1939). This painting, made by the Mexican artist after her divorce from her husband, Diego Rivera, represents her as a dissociated woman with a dual personality and two hearts, one healthy and the other exposed, both united by a vein but with a severed circulatory system. She is emotionally bleeding. The painting does not clearly express her pain, however, because she has taken an emotional distance from her own pain. Montano, similarly, points to a "dissociative traumatic subjectivity" and puts her mind and her heart into communication, even though in moments of disruption, they have seemed completely disconnected.

Montano's blending of heartbeats, heartbreaks, performance, and therapy finds an antecedent in Barbara T. Smith's *Ritual Meal* (1969). Conceived after a traumatic divorce, for the six-course dinner performance, guests dressed in medical scrubs ate raw meat with surgical instruments while images of a surgically extracted beating heart were projected overhead. The image of the heart and the serving of raw meat point to a cannibalistic ritual through which Smith hoped to heal her own heart. She was offering it symbolically at a dinner as a way of recovering power and energy and as a rite of initiation to her new life as an artist.[58]

Another major influence for Montano was Pauline Oliveros, most particularly her *Sonic Meditations* (1971) (discussed in chapter 2), in which she mentions healing and the active listening to the heart.[59] Her Deep Listening exercises include sensing heartbeats by touching and then clapping and walking with a group, making rhythm cycles,[60] and analyzing its tempo in relation to other surrounding rhythms[61] or opening the heart chakra with chi kung.[62] Listening to the heart for healing became an important practice for Oliveros throughout her whole career. These conceptions of healing and regeneration of the self through listening to one's heart and the connections between sounds that produced an awareness of internal and external worlds that expand a sense of consciousness are an interest shared by both Oliveros and Montano.

Montano further explores the spiritual and meditative locus of the heart in *7 Years of Living Art* (1984–1991), which later became *14 Years of Living Art* (1984–1998). Executed mostly in her house, where the artist lived in a meditative state, the self-discipline or imposed "art vow" combined art and life activities, in which each year corresponded to one of the seven

Hindu chakras. As Jennifer Fisher has noted, "Throughout the seven years, a time period in which each cell in the physical body is regenerated, Montano tracked the changes in her body."[63] Using an individual interpretation of yogic practices, she used the energies of the *chakras* to access intuitive knowledge, assigning each of the seven chakras to daily life experiences and to each year of the seven-year cycle. Montano matched each chakra with a location, a color, a tone, and a state. The *anahata* chakra or heart center—which is associated with calmness, balance, and serenity—is in the heart, and Montano linked it to the color green, the musical note *A*, and compassion.

Montano engaged with other people's hearts as part of this lengthy piece. During her *Art/Life Counseling*, she conducted seven-hour-long advice-giving and tarot and palm readings sessions each month in a window of the New Museum in New York. To this day, she has continued this practice of using her art to create a confessional space in which she invites people to open their hearts and say whatever is on their minds.

Montano has also embraced the metaphorical and practical connotations of heartbeats in her writings, which she uses as a therapeutic or psychoanalytic personal dialogue for clarifying thoughts and dealing with emotions. In her book *Letters from Linda M. Montano* (2005), she frequently refers to her heart and the use of performance "to open the heart" in order for participants to "learn hidden information about themselves." In her text "Death," the heart becomes a source of therapy for mourning: "Listen to your heart. What does it say about death? Listen to each room of the heart: room 1, room 2, room 3, room 4. Give your heart attention."[64] In "Portrait of Sappho," Montano considers the heart as the source of life and as something that can be nostalgically remembered by a death: "My heart was a muscle the size of a grapefruit / Once, my heart pumped 766,000 gallons of blood every year / Once, my heart . . . / Once, my heart . . . / My heart . . . / My heart . . ."[65]

Since the artist reconverted to Catholicism, Montano has also created various performances depicting the lives of female Catholic saints. *Teresa de Ávila* (2007) focuses on the sixteenth-century Spanish saint, mystic, anorexic, and writer who recounts a "transverberation" or mystical vision while at prayer in which an angel repeatedly pierced her beating heart with an arrow.[66] Subsequently, Teresa is often depicted in ecstasy, as in

Bernini's famous sculpture of her next to a Sacred Heart of Jesus. This wounded, flaming heart pulsates light and represents the love of God. The cult of the Sacred Heart of Jesus gained momentum in the seventeenth century through the visions of Marguerite Marie Alacoque, which led to the formal recognition of the cult by the Vatican in the mid-nineteenth century by petition of Mary of the Divine Heart. In *The Essence of Christianity* (1841), philosopher Ludwig Feuerbach, reflecting the centrality of the heart in medical and religious discourses of the nineteenth century, compares religious feelings with the vital movements of the heart:

> As the action of the arteries drives the blood into the extremities, and the action of the veins brings it back again, the life in general consists in a perpetual systole and diastole; so is it in religion. In the religious systole man propels his own nature from himself, he throws himself outward; in the religious diastole he receives the rejected nature into his heart again.[67]

More than a century later, Luce Irigaray, a feminist philosopher whose work muses on religion and spirituality, when addressing the "sensible transcendental," she fuses the mind and the body, the divine and the carnal, when she declares "the sensible remains sweet and carnal: living, pulsating."[68] Montano's performance as Teresa de Avila explores themes of faith, devotion, and religious sentiments. Channeling the energy and spirit of Teresa, the artist asks God to heal her wounded heart. The performance is a form of medicine and a way of reaching the states of compassion, understanding, and wisdom that mystics achieve, using the heartbeat as a vehicle of spiritual connection.

For Montano, heartbeats are "the voice of everything," representing life and love but also death and loss. All of these factors are present in *The Psalms R Us* (2018), a more recent performance, which included components of some of her past works: therapeutic confessions from participants, reading aloud of psalms, her dressing as a nurse, and the use of a stethoscope. Created after being hospitalized for high blood pressure and anxiety, Montano turned to listening to the heartbeats of the performers who were collaborating with her. They opened their hearts publicly and shared emotional memories while Montano acted as a soul nurse, monitoring how laughing and crying altered their pulse rates.[69]

Heartbeats have been a constant in Montano's work. As an artist interested in life, death, and religion, she has made work with all her heart and has made her art and life heart-beating objects of healing through meditation, sacrifice, and communal experience. Heartbeats are an invisible language, the language of the heart, that organ that dissolves the boundaries between the self and the other, that little muscle that always endures and has the capability to silence the mind and facilitate healing. And today, when eighty-year-old Linda Montano monitors her blood pressure with a cuff to check her palpitations, her heart murmurs daily: "LOVE, LOVE, LOVE . . . HAVE A LOVING HEART. THAT IS THE HIGHEST ART."[70]

(IM)PERMANENCE: CHRISTIAN BOLTANSKI

Upon entering Christian Boltanski's *Personnes* (2010; figure 7.4), you hear thousands of beating hearts resounding throughout the Grand Palais in Paris. Beyond a large wall made of urn-like tin boxes, you find a landscape made of tons of discarded clothing distributed in a mountainous heap and a grid of sixty-nine rectangles that stretched throughout the main nave of the exhibition space. The heartbeat sounds come from loudspeakers carefully placed on steel poles throughout the gallery. This cardiac symphony is composed of recordings that visitors have donated willingly at a separate booth. Heartbeats and old clothes function as a metonymic identification not only of human life but also of a death industry represented by a large mechanical claw that repeatedly digs into the mountain of clothing. "[It] picks up clothes by chance, in the same way we fall by chance," Boltanski said.[71] In this industrial and cold environment, permeated by a sense of mortality and mourning, the heartbeat is a material that preserves a presence in absence. Its ominous and factory-like rhythm reinforces the serial, dramatic, and ghostly spirit of the installation, irremediably evoking the beating hearts of people who suffered in concentration camps, those who suffer in ongoing wars, as well as those who dwell in refugee camps today.

Born in Paris in 1944, the year in which Western Europe was liberated from the Nazi regime, Christian Boltanski's early life was haunted by war

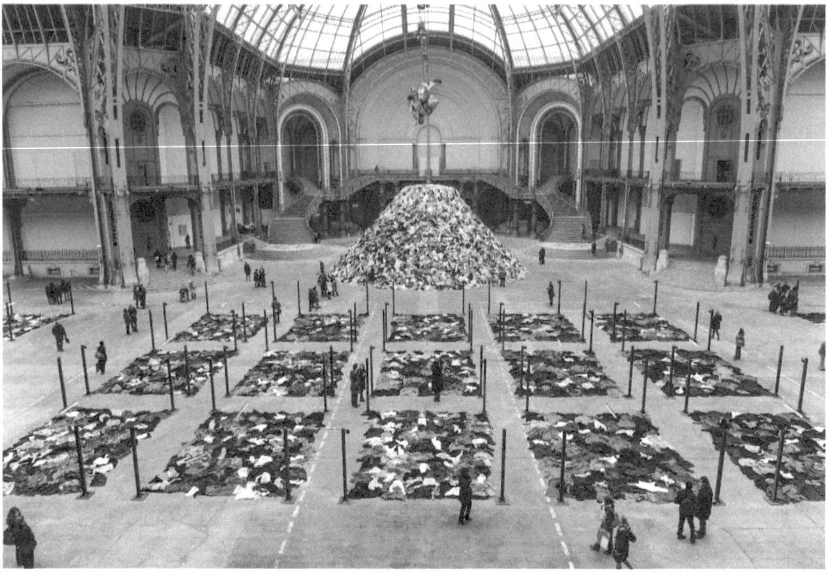

trauma and family memories of the reality of the Holocaust. For two years, his father, a Ukraine-born Jewish doctor, was forced to hide, with a racing heartbeat, under the floorboards of his family's apartment to avoid discovery and death. Being raised amid the remains of so much trauma made Boltanski comfortable with death and how survivors live on. Until his death in 2021, he produced an oeuvre that used debris from the everyday to explore life, death, and what he referred to as "small memory," or those things that are easily forgotten and are not to be found in books.[72] Being an artist for Boltanski was about "capturing life, life that has gone, or life as it exists."[73] Heartbeats feature in several of his works as a way

7.4

Christian Boltanski, *Personnes*, 2010. Monumenta, 2010, Grand Palais, Paris. Courtesy Fonds de dotation Christian Boltanski & Marian Goodman Gallery. © and Courtesy Fonds de dotation Christian Boltanski, ADAGP.

to capture the evanescent and reflect on life's multiplicity, regeneration, and finitude.

As a self-described "sentimental minimalist," Boltanski created striking memorial works that speak to the tragedy of death, the loss of humanity, the lives lost in the Shoah, and the impossibility of remembering the myriad human lives that have preceded us. To learn something by heart is to memorize it or, in Boltanski's case, to memorialize it. The French artist devoted his life to creating "monuments and memorials using as medium the human trace and the memento mori."[74] His works frequently include collections that serve as indexical traces of the past, such as his heartbeat recordings in *Les Archives du Coeur* (2008–) or his collection of Holocaust-evoking portraits in *The Children of Dijon* (1986). Through these assemblages, he sought to make us recognize that "we cannot make people live again, nor fight against death, but we can say: 'There was.'"[75]

Boltanski's heartbeat works are archival, past-focused, introspective, sober, mournful, poignant, and deeply moving. This contrasts with Lozano-Hemmer's pulse works, which are anti-archival, present-focused, communal, joyful, dialogic, electrifying, and vivid. And while Yoko Ono's *Heartbeats/Syria* raise awareness of the heinous atrocities and lives lost at a given time and location, Boltanski's works are all-embracing and referring to various temporalities, evoking past, present, and future lives and deaths. In fact, for Boltanski, the beauty of working with heartbeats was that they come from a real person, from someone who loves, loved, or was loved. Cardiac recordings and clothes are signs of a missing subject; they are a naked testament to a present and past life. This is always to be seen in stark contrast to the Nazis, who controlled the biological, as well as life and death, and treated humans as disposable objects

> [In] the heartbeats: there was somebody. There was somebody, and now we have an object, but we can destroy it if we want, because it's only an object. And if you speak about the Holocaust, there's always this idea with the Nazis that they were dealing not with people but with objects. The Nazis always said, "Today we have 20 tons of material arriving." . . . That was the terminology.[76]

Boltanski asserts that there are more instances of this worrisome objectification: "Horror is always there. People ask, why did the people who knew about the Shoah, and about Stalin, refuse to speak? But we know about Syria, about migrants. And what are you doing? What am I doing? Nothing."[77]

Personnes was purposely designed as an immersive experience that created a strong sense of oppression. The only comfort comes from getting outside of that throbbing environment, away from that space of encounter with the death of others that also confronts us with our own mortality. As noted by Jean-Luc Nancy, there is a togetherness in the idea of death because it always relates to the being and the death of others.[78] We can only become aware of our own temporary presence in this world through the disappearance of others and the reunion with what once was. For Nancy, death is unable to be represented since it is only ever represented for those who remain. It is in fact impossible to represent the twentieth-century genocides and the current atrocities that continue to happen today. But what matters is not its representation but the image of its irrepresentability, of its ghastliness, and that is what Boltanski achieved with his heartbeat work. As death itself cannot be fully portrayed because it is an abstraction, it can only be mourned and invoked through its traces.

Heartbeats are a significant trace since they represent the beginning of life, life passing, and the end of life. As a trace incarnating what is permanently lost, surviving the death of people, heartbeats are a sound of life but also of fragility, transience, and mortality. Upon leaving *Personnes*, visitors to the Grand Palais could record their heartbeats assisted by white robe-dressed technicians using a mic-equipped stethoscope. "I am leaving a part of myself in this work," a visitor declared, "This relationship between life and death is troubling. But this work allows you to think about life, to continue to live."[79] Creating a deep sense of absence with their presence, these recordings are subject and object, sign and index, a reminder of the imminence of death, the passing of time, the horrors of mass murder but also a reinvigorating *carpe diem*.

Boltanski's obsession with heartbeats, which he conceived as "ready-mades" and a reflection of our restlessness, dates back to 2005, when he recorded his own cardiac rhythm for *Le Coeur* (2005). Presented for the first time as part of his solo show *Prendre la Parole* at Marian Goodman Gallery in Paris, the installation consisted of a dark and narrow space with a single bulb flickering to the rhythm of the artist's heart sounds.[80] Recalling the piece, Boltanski declared,

> Light in itself can be very unpleasant, just as with sound. Some people enjoy spending time in the *Le Coeur* room; others can barely stand it. It's because you always get the feeling that the heart will stop beating at any moment. I felt pretty uneasy myself by the time I'd finished working on it. I went straight along to see a cardiologist. I had the impression my heart wasn't beating properly. It's strange to hear something from your own insides—you always get the feeling that something's wrong.[81]

After the creation of *Le Coeur*, Boltanski decided to develop a heartbeats album as a way in which one could once again listen to loved ones.[82] *Les Archives du Coeur* (2008–; figure 7.5) is his collection of thousands of human heartbeats retrieved from people around the world. Since 2010, it has been permanently located in a purpose-built museum located on the remote Japanese island of Teshima. In this refuge, poetically conserving the lives of thousands of people through carefully arranged and labeled heartbeat recordings, Boltanski's own heartbeat blinks against death and oblivion in the company of nearly a million others. Boltanski has given these traces of anonymous death a new life: "It's like a kind of resurrection."[83] All heartbeats, like all portraits, speak of a past moment that will never return, a person or memory that we can never be close to again. By creating a collection of heartbeat recordings that could repeat over time, Boltanski intends to create a form of immortalization via digital stethoscopes, sensors, computers, and recording technologies, a chimeric memorial that offers a notion of posterity: a palliative to mortality.

Les Archives du Coeur (figure 7.6) was purposely designed to create an emotional impact through the play of light, images, and sound. When you enter the three-room building, a computer screen provides the name and the place where the heartbeats currently being played in the adjoining Heart Room were recorded. There, a single light bulb illuminates the dark space, pulsating to the rhythm of someone's heartbeat, a sound that feels eerie. The walls of the room evoke a columbarium, covered in black portrait-format boards of various sizes, in which you can only see the reflection of your own face and the blinking lights. Your image is there,

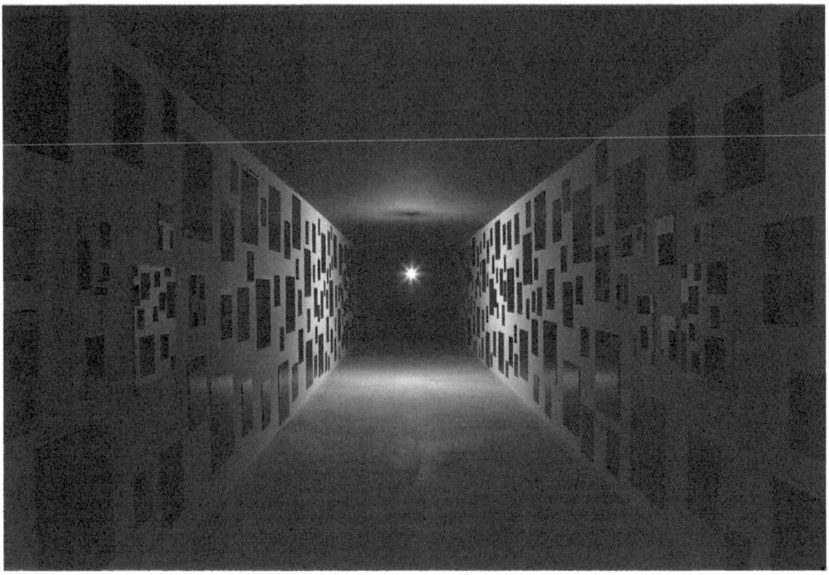

in front of the object, but without your presence, it disappears, leaving no memory. In the final room, overlooking the sea, you can consult the database of heartbeat audio files and make a future relic of yourself by recording your own heartbeats using a stethoscope with a microphone connected to a computer. At the end of the process, you receive a copy of the audio track on compact disc. Although all these heartbeats seem to sound similar to each other, there is a sense of uniqueness to each because, as Boltanski said, "there would always be something individual within the collective."[84]

Les Archives du Coeur is not about the space itself, however, or its objects. Instead, it aims to create a mythic work, one that transcends its

7.5

Christian Boltanski, *Les Archives du Coeur*, 2010, Teshima, Japan. Photo by Kuge Yasuhide. Courtesy of Fukutake Foundation.

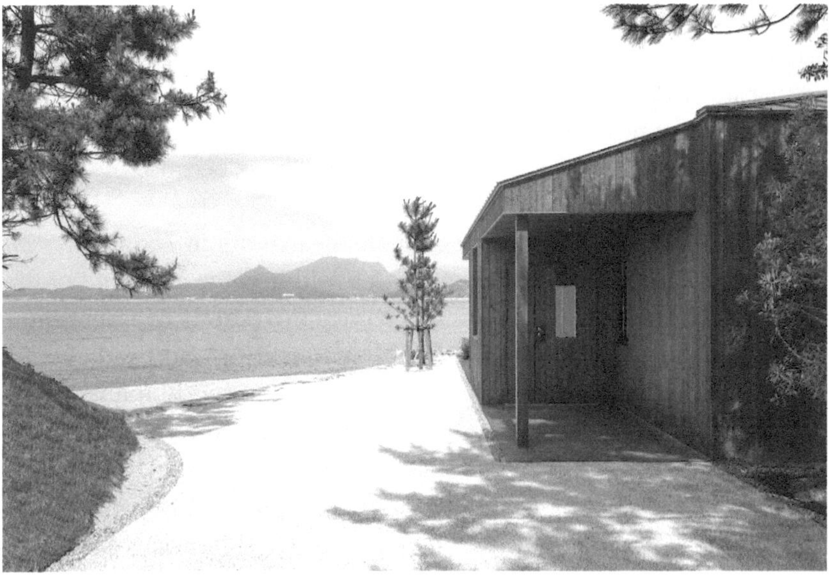

own physicality: "What I wish but will never happen is to have all the heartbeats on earth, a Utopian ideal and totally impossible, of course,"[85] Boltanski declared. "I'm interested in the creation of myths; you don't have to go to Japan to know that the artwork exists. Simply knowing that the artwork exists is what makes it exist as an artwork."[86] The idea that the heartbeats of thousands of humans are preserved on a distant island is the most important aspect of the work. It is its imaginary and symbolic order that produces a story.

Early in his career, Boltanski, in his artist book *Recherche et presentation* (1996), considered his collection of traces and archival structures as a form of survival, a victory over death. Perhaps through "a big collective

7.6

Christian Boltanski, *Les Archives du Coeur*, 2010, Teshima, Japan. Photo by Kuge Yasuhide. Courtesy of Fukutake Foundation.

effort" of self-preservation, this would be more effective than the efforts of doctors and scientists who "merely establish a pact" with death by postponing it.[87] Decades later, though, he declared, "From the beginning I knew it was not possible, only a kind of dream . . . Nothing is preserved, I've always said my work has always missed the boat, because you cannot fight against forgetting."[88]

With the elapse of time, most of the heartbeats in *Les Archives du Coeur* will be those of dead people, and it will become an island of the dead. Indeed, Boltanski's heartbeat recorded in Teshima in 2010, is now in the company of thousands of heartbeats of living and dead people. Boltanski regarded death as a natural occurrence that humans always try to avoid: "Death has become a thing to be ashamed of. Being old too. Death, however, is eminently natural: as soon as we're born, we're going to die."[89] French historian Phillipe Ariès, in his book *Western Attitudes Toward Death*, referred to this ambiguity by noting that "technically, we admit that we might die. . . . But really, at heart we feel we are nonmortals."[90] This desire for survival and immortality, as Nancy noted, is part of a cultural program that seeks the control of nature.[91] In response, people often resort to forms of representation, such as heartbeat recordings, in order to retain a semblance of presence. To account for this tendency, Boltanski recalled a story:

> Naturally, all heartbeats are much the same but, at the same time, they change according to moments in one's life. Mine today is not the same as five years ago. Very recently a woman rang me saying: "My sister's dying, do you want to record her heartbeat?" I refused. I'm not a doctor. I couldn't see myself going to the hospital to record her [sister's] heart. I explained to her how to do it herself, as it's really easy. The woman had the feeling that if I recorded her sister's heartbeat, something of it would remain. In fact, nothing would remain but a relic.[92]

Intentionally contradictory, mirroring human life's paradoxes, Boltanski's oeuvre presents, on the one hand, heartbeats as a means of fighting against forgetting and as a form of resurrection. But on the other, he denies the value of relics: "You can hold onto the clothes, and even the heartbeats of many, many people," he said, "but you can't keep anybody."[93] As the curator and art historian Catherine Grenier has observed,

Boltanski's oeuvre is "designed to halt death, to freeze life forever, and, at the same time, dedicated to express[ing] the ever-present consciousness of extinction—such is the paradox assumed by Boltanski since he started out as an artist."[94] This ambiguity is at the core of his work and reflects the inherent binary contradictions of life and death.

Boltanski's art is not about preserving life; rather, it emphasizes its absence, the lives lost. By playing with emotionally charged materials, it explores the drama of disappearance and mourning. Heartbeats are faceless portraits that transform absence into powerful presence, they are like photographs, like garments, the trace of a life. As Walter Benjamin noted, photographs and recordings are part of "the cult of remembrance of loved ones, absent or dead," the evidence of life from which an aura emanates from the expression of a human face or voice.[95] Since heartbeats represent the voice of the heart, Boltanski wants to achieve the impossibility of immortality by creating a memorial island that offers a sense of posterity but is mostly a place to contemplate the inevitability of death and oblivion. "It doesn't matter if you're rich or poor, good or evil, we're fundamentally the same and we will all be forgotten—that's the truly radical nature of transience,"[96] Boltanski said. "It doesn't matter how many archives you set up—it's just not possible to hold on to life."[97]

Certainly, nothing resists death; each heartbeat is an advance toward it, but a community of achieved heart rhythms can make us aware that we live in the same world and that our hearts beat together. This can give an individual the unique sensation that there is never a single being without the existence of another being, just as there is no life, and life is no life without the existence of another heartbeat.

Life is a rhythmic recurrence of pulsating finitude.

CONCLUSION

Walking on the streets in Berlin, at a time when the Russian invasion of Ukraine and Israel's atrocities in Palestine are dominating the headlines, I found graffiti of an anatomical heart with a background of multicolored zigzag lines. In a corner of the wall, it read, "Many colors. Different ways. Same Source." That image, that phrase, aligns with the essence of this book. In this manuscript, I have shown how many artists from different locations and generations have used one single source, the heartbeat, in very various and significant ways: sometimes to seek empathy toward war victims, to make an awareness of life, and to make us realize that we are all equal and interconnected, sharing a common physical and energetic bond.

Heartbeats are a vital sign in revealing the history of inquiring about being alive. I have traced in this book a lineage of heartbeats in the arts: from ancient visions to the creation of instruments and the expansion and appropriation of cardiac recording technologies by artists and composers, who show the heterogeneous and multiontological spirit of the heartbeat. The intention of this book has been, on the one hand, to demonstrate how artists, aided by different techniques and technologies, whose origins stretch back to the nineteenth century, have found a myriad of uses for heartbeats, turning a private inner movement into a social element. They have all transformed instruments intended to create distance into tools for reflection and connection. On the other, this book has also aimed to expand our understanding of heartbeats through their use in the arts, moving beyond the conception of a simple physiological phenomena toward something more heterogeneous, rich in multiple meanings, and connotations. The findings that I have presented suggest that the heartbeat in these works is a polysemous, symbolic, aesthetic, and metaphoric sign of material and immaterial meanings that bridges aspects of ontology, technology, biology, epistemology, history, art, science, corporality, literature, diverse social realities, organizations, and environments. The heartbeat consistently complicates demarcations of feminine/masculine, self/other, inside/outside, sign/index, and even death/life because heartbeats are not only a physical phenomenon but also time, memory, transferable energy, a form of portraiture, information, closeness, a healing source, etc. The heartbeat is a sign that is both physical and spiritual, mysterious and mundane, private and public, mechanical and emotional, passive and active, personal but alien, palpated but inapprehensible, intimate and public, sometimes steady while at other times difficult to control.

The heart beats with its electromagnetic pulsing power, a source that connects us not only to other humans but also to other living creatures, to the pulsing Earth and the cosmos. This is something that indigenous cultures have known since ancient times through their cosmologies and technologies. We are all interconnected and interdependent, and the heartbeat of the human heart connects with the heartbeat of the Earth and those found in the cosmos. In her work, artist Diana Domingues merges technology and indigenous Amazonian rituality in a way that reminds us that spiritual energy, technological energy, and corporeal energy are linked. Also, Pauline Oliveros and Éliane Radigue in their compositions connect the heartbeat to other sounds of the environment and the universe.

Medical instruments, such as the stethoscope and EKG, have their own agency when they are deployed by artists. Artists show how the heart is more than organic matter and electrical phenomena but something with multiple ontologies and epistemologies. As I have shown, heartbeats can tell us about the gendering of the art world, medicine, technology, and cardiac activity. Some artists expose gender inequalities in both the arts and sciences. In fact, men in the science and arts, as we have seen, had earlier access to medical technologies. However, female artists were able to overcome obstacles, sexism, and gender biases, and created extremely compelling and pioneering works using medical technologies that address emotional, philosophical, ritualistic, technological, and sociopolitical concerns. Although science considers the heartbeat as something mechanical (i.e., a pump or a beeping), and although it can be something collected, stored and interpreted by governments and enterprises, the heartbeat is a connective embodied emotional language in which the ontological and epistemological, the biological and technological, and the individual and the social, the rational and the emotional, and nature and culture are intertwined.

The work of the artists addressed in this book reveals that our hearts are all the same, depending on the constant beating of a heart, which is red when filled with blood, white when dead, and transferable from one body to another, no matter race, class, or age. We are the same matter, and the heartbeat interconnects us and connects us to the universe with its vibrational, electrical, and emotional faculties. Heartbeats are connecting ecosystems that have a variety of connotations that facilitate multiple typological interactions and aesthetic engagements. The multiplicity of layers that were unfolded in this book mapped different interpretations of core forms detected—measuring, space, electrical, time, transplants, surveillance, and regenerations—that reflect the heartbeat as a complex and heterogeneous sign, entangled in the quadrants of technology, biopolitics, labor, migration, care, and control uses. Following this vein, we can say that technoscience has a double character, which both portrays and records, and also liberates and enslaves.

From the liberating point of view, I have shown through the work of Yoko Ono, Linda Montano, and Christian Boltanski how the heartbeat has been a source of therapy and meditation and connects us with people of our present or with future generations. The rhythm of the heart has served as a source of meditation that influences the emotional, mental, and physical state of a person. This makes them a source of empathy for others in more difficult and distant situations, such as victims of crime, war, or displacement, whose fragile lives hang by a thread as thin as a heart trace.

Heartbeat works are the condensation of centuries of experimentation and scientific knowledge, and despite containing theories, proven and verifiable techniques, and scientific postulates, they force us to ask questions intrinsic to human existence, mainly about the relationship between humans, their technological environments, and nature.

I have revealed that narratives of the heartbeat works and their histories are interdisciplinary, connecting different fields and notions.

The universe of heartbeats works is vast and has moved in tune with the development of technologies, history, and sociopolitical phenomena. With the rise of new digital and surveillance technologies, such as AI, in the twenty-first century, heartbeats continue to generate new possibilities for artists to experiment with their own and other's vital signs. Without a doubt, the use of heartbeats in art will be a source of inspiration as long as the soft lub-dub sound of a heart persists.

NOTES

INTRODUCTION

1. Mrinal Kanti Das, Soumitra Kumar, Pradip Kumar Deb, and Sundeep Mishra, "History of Cardiology in India," *Indian Heart Journal* 67, no. 2 (2015): 164.
2. E. A. Wallis Budge, *The Book of the Dead* (New York: Gramercy Books, 1995).
3. Bin Liu, "Integrative Views of the Heart in Chinese and Western Medicine," *Integrative Medicine International* (2017): 46–51.
4. In fact, the heart's word *yóllotl* comes from the word *yollo*, which means "what moves." Carlos Viesca et al., "El corazón y sus enfermedades en la cultura Náhuatl Prehispánica" *Estudios de Cultura Náhuatl* 36 (2005): 229.
5. Miguel León-Portilla, "Significados del corazón en el México Prehispánico" *Estudios de Cultura Náhuatl* 35 (2004): 252; Mónica Chávez Guzmán, *Cuerpo, enfermedad y medicina en la cosmología maya del Yucatán colonial* (Mexico City: Universidad Nacional Autónoma de México, Centro Peninsular en Humanidades y Ciencias Sociales, 2013), 73–78.
6. For more on Aristotle, see Charles G. Gross, "Aristotle on the Brain," *The Neuroscientist* 1, no. 4 (1995): 245–250.
7. Ecole Polytechnique Fédérale de Lausanne, "Your Brain Suppresses Perception Related of Heartbeat, for Our Own Good," May 4, 2016. https://actu.epfl.ch/news/your-brain-suppresses-perception-of-heartbeat-for-/.
8. Nancy G. Siraisi, "The Music of Pulse in the Writings of Italian Academic Physicians (Fourteenth and Fifteenth Centuries)," *Speculum* l (1975): 689–710.
9. Claudia Arozqueta, "Heartbeats and the Arts: A Historical Connection" *Leonardo* 51, no. 1 (2018): 35. Text received in 2014.
10. Evan Bedford, "The Ancient Art of Feeling the Pulse," *British Heart Journal* 13, no. 4 (1951): 423.
11. Shigehisa Kuriyama, *The Expressiveness of the Body and the Divergence of Greek and Chinese Medicine* (New York: Zone Books, 1999), 5.
12. Kirstie Blair, *Victorian Poetry and the Culture of the Heart* (Oxford: Oxford University Press, 2006), chapter 2, https://www.doi.org/10.1093/acprof:oso/9780199273942.001.0001.
13. Fay Bound Alberti, *Matters of the Heart: History, Medicine and Emotion* (Oxford: Oxford University Press, 2006).
14. Lynda Birke, *Feminism and the Biological Body* (Edinburgh: Edinburgh University Press, 1999).

15. Anne Pollock, "Heart Feminism," *Catalyst: Feminism, Theory, Technoscience* 1 (2015): 1–30.
16. James Peto, ed., *The Heart* (New Haven: Yale University Press; London: Wellcome Collection, 2007), xv.
17. The artist analyzed were John Cage, James Turrell, James Tenney, and William Burroughs. Douglas Kahn, "Let Me Hear My Body Talk, My Body Talk," in *Relive: Media Art Histories*, ed. Sean Cubitt and Paul Thomas (Cambridge, MA: MIT Press, 2013), 235–256.
18. Sean Cubitt and Paul Thomas, "Introduction: The New Materialism in Media Art History," in *Relive: Media Art Histories*, ed. Sean Cubitt and Paul Thomas (London: MIT Press, 2013), 6.

CHAPTER 1

1. Evan Bedford, "The Ancient Art of Feeling the Pulse," *British Heart Journal* 13, no. 4 (1951): 423.
2. Shigehisa Kuriyama, *The Expressiveness of the Body and the Divergence of Greek and Chinese Medicine* (New York: Zone Books, 1999), 32.
3. Bendix Ebbell, *The Papyrus Ebers: The Greatest Egyptian Medical Document* (London: Oxford University Press, 1937), 115.
4. The word "artery" in Greek derived from the word "*αηρ*," air.
5. Fritz Steckerl, *The Fragments of Praxagoras of Cos and His School* (Leiden: E. J. Brill, 1958), 25.
6. Heinrich von Staden, *Herophilus: The Art of Medicine in Early Alexandria* (Cambridge: Cambridge University Press, 1989), 666.
7. Kuriyama, *The Expressiveness of the Body*, 32.
8. Claudia Arozqueta, "Heartbeats and the Arts: A Historical Connection" *Leonardo* 51, no. 1 (2018): 33.
9. Nima Ghasemzadeh and A. Maziar Zafari, "A Brief Journey into the History of the Arterial Pulse," *Cardiology Research and Practice*, vol. 2011, Article ID 164832, 2011, https://doi.org/10.4061/2011/164832.
10. William Osler, *The Evolution of Modern Medicine* (Champaign, IL: Project Gutenberg, n.d.), 55, accessed October 3, 2016, https://www.gutenberg.org/files/1566/1566-h/1566-h.htm.
11. More information about Galen's theories can be found in Michael Boylan, "Galen: On Blood, the Pulse and the Arteries," *Journal of the History of Biology* 40, no. 2 (2007): 207–230.
12. Kuriyama, *Expressiveness of the Body*, 34.
13. Boylan, *Galen*, 221.
14. Mohammed Chamsi-Pasha and Hassan Chamsi-Pasha, "Avicenna's Contribution to Cardiology," *Avicenna Journal of Medicine* 4, no. 1 (2014): 9.

15. Francis Wells, "The Renaissance Heart. The Drawings of Leonardo da Vinci," in *The Heart*, ed. James Peto (New Haven, CT: Yale University Press; London: Wellcome Collection, 2007), 83.

16. Bryan Mowry, "From Galen's Theory to William Harvey's Theory: A Case Study in the Rationality of Scientific Theory Change," *Studies in the History and Philosophy of Science* 16, no. 1 (1985): 49–82.

17. A theory that survived into the seventeenth century, the imbalance of humours or *dyscrasia* was produced by an altered interaction of the four bodily fluids (blood, black bile, yellow bile, and phlegm), which also defined temperaments and emotions. High blood or phlegm levels were thought to make a person sanguine or phlegmatic and prone to lovesickness; an excess of yellow bile was present in choler persons prone to anger; and those with high black bile levels were melancholic and prone to sadness. Alberti, *Matters of the Heart*, 19.

18. L. S. Dixon, *Perilous Chastity: Women and Illness in Pre-Enlightenment Art and Medicine* (Ithaca: Cornell University Press), 11.

19. Michel Foucault, *The History of Sexuality, Volume I: An Introduction* (New York: Pantheon Books, 1978), 145.

20. Archibald Billing, "Clinical Lecture," *The Lancet: A Journal of British and Foreign Medicine* 1 (1832): 426.

21. Gary L. Townsend, "Sir John Floyer (1649–1734) and His Study of Pulse and Respiration," *Journal of the History of Medicine and Allied Sciences* 22, no. 3 (July, 1967): 302.

22. D. D. Gibbs, "The Physician's Pulse Watch," *Medical History* 15, no. 2 (1971): 187–190.

23. Audrey B. Davis, *Medicine and Its Technology: An Introduction to the History of Medical Instrumentation* (London: Greenwood Press, 1981), 88.

24. René-Théophile-Hyacinthe Laennec, *A Treatise on Mediate Auscultation and on Diseases of the Lungs and the Heart* (London: Bailliere, 1846), 29. In this particular instance, the sound was probably caused by respiration.

25. Laennec, *A Treatise on Mediate Auscultation*, 2.

26. Peter Pesic, "Music, Mechanism, and the 'Sonic Turn' in *Physical Diagnosis*," *Journal of the History of Medicine and Allied Science* 71, no. 2 (2016): 161.

27. Michael J. K. Bokor. "When the Drum Speaks: The Rhetoric of Motion, Emotion, and Action in African Societies," *Rhetorica: A Journal of the History of Rhetoric* 32, no. 2 (2014): 165–194; Tanya Y. Price, "Rhythms of Culture: Djembe and African Memory in African-American Cultural Traditions," *Black Music Research Journal* 33, no. 2 (2013): 227–247.

28. Miguel León Portilla, "La música en el universo Náhuatl," *Estudios De Cultura Náhuatl* 38 (2007): 142.

29. Elemire Zolla, "Korean Shamanism," *RES: Anthropology and Aesthetics* no. 9 (1985): 101–113; János Kárpáti, "Music of Female Shamans in Japan," *Studia Musicologica* 54, no. 3 (2013): 225–256.

30. Emma Borjigid, "The Shamanic Healing Power of the Drum" (unpublished manuscript, 2017), PDF, 4–5.
31. Ibrahim M. Eltorai, *A Spotlight on the History of Ancient Egyptian Medicine* (Boca Raton, FL: CRC Press, 2020), 126.
32. Jonathan Sterne, *The Audible Past* (Durham, NC: Duke University, 2003), 118.
33. J. James Smith, "The Inventum Novum of Joseph Leopold Auenbrugger," *Bulletin of the New York Academy of Medicine* 38, no. 10 (1962): 698.
34. Hillel Schwartz, *Making Noise: From Babel to the Big Bang & Beyond* (New York: Zone Books, 2011), 210.
35. Laennec, *Treatise on Mediate Auscultation*, 4.
36. Laennec, 5.
37. Authors like Jacalyn M. Duffin, Jonathan Sterne, and Stanley Joel Reiser have put special emphasis on the fact that mediate auscultation transformed the doctor's status in regard to the patient. See Sterne, *The Audible Past*; Jacalyn M. Duffin, "The Cardiology of R. T. H. Laennec," *Medical History* 33, no. 1 (1989): 41–71; and Stanley Joel Reiser, *Medicine and the Reign of Technology* (Cambridge: Cambridge University Press, 1978), 12.
38. Maxwell David Melin, "The Industrial Revolution and the Advent of Modern Surgery," *Intersect* 9, no. 2 (2016): 3.
39. Jonathan Sterne, "Mediate Auscultation, the Stethoscope, and the 'Autopsy of the Living': Medicine's Acoustic Culture," *Journal of Medical Humanities* 22, no. 2 (2001): 135.
40. Bokor, "When the Drum Speaks," 171.
41. Suzy Leinster, "Indigenous Drumming Has healing Powers," *The Queen's University Journal*, September 26, 2023, https://www.queensjournal.ca/indigenous-drumming-has-healing-powers/.
42. Bedford, "The Ancient Art of Feeling the Pulse," 426.
43. Kuriyama, *Expressiveness of the Body*, 85.
44. Nancy G. Siraisi, "The Music of Pulse in the Writings of Italian Academic Physicians (Fourteenth and Fifteenth Centuries)," *Speculum* l (1975): 689–710.
45. Siraisi, "The Music of Pulse," 693.
46. Leonardo Da Vinci, *Leonardo on the Human Body* (New York: Dover Publications, 1983), 100.
47. Kuriyama, *Expressiveness of the Body*, 81.
48. Pesic, "Music, Mechanism, and the 'Sonic Turn," 148.
49. Sterne, "Mediate Auscultation," 123.
50. Schwartz, *Making Noise*, 211.
51. Laennec, *Treatise on Mediate Auscultation*, 479.
52. Laennec, 567.
53. Laennec, 566–567.

54. Laennec, 544.

55. Davis, *Medicine and Its Technology*, 91.

56. Sir John Forbes, preface to René Laennec, *A Treatise on the Diseases of the Chest* (London: T. and C. Underwood, 1821), xviii–xix.

57. Translated by the author. The original French: "Au moyen d'un tube appliqué aux parties extérieures du corps, notre savant compatriote breton, le docteur Laënnec, est parvenu à reconnoître, par la nature du bruit de la respiration, la nature des affections du coeur et de la poitrine. Cette belle et grande découverte fera époque dans l'histoire de l'art. Si l'on pouvoit inventer une machine pour entendre ce qui passe dans la conscience des homes, cela seroit bien utile dans les temps où nous vivons." Françoise-René de Chateaubriand, *Oeuvres Complètes. Tome Dix-Huitième* (Paris: Pourrat Frères Editeurs, 1836), 327.

58. Melissa Dickson, "Thou Simple Tube," *History Today* 67, no. 5, (2017): 66–77.

59. Some of these anxieties are documented in Dickson, "Thou Simple Tube."

60. Kirstie Blair, *Victorian Poetry and the Culture of the Heart* (Oxford: Oxford University Press, 2006), 26.

61. Blair, *Victorian Poetry*, 23.

62. Blair, 60.

63. Claude Richard, "The Heart of Poe and the Rhythmics of the Poems," in *Critical Essays on Edgar Allan Poe*, ed. Eric W. Carlson (Boston: G. K. Hall & Co., 1987), 195–206.

64. Edgar Allan Poe, *The Tell-Tale Heart* (Champaign, IL: Project Gutenberg, n.d.).

65. Davis, *Medicine and Its Technology*, 109.

66. J. M. Forrester, "An Experiment of Galen's Repeated" *Journal of the Royal Society of Medicine* 47, no. 4 (1954): 242.

67. Emil Vollmer, *Leonardo Da Vinci* (New York: Reynal, 1960), 383.

68. M. E. Silverman, "William Harvey and the Discovery of the Circulation of Blood," *Clinical Cardiology* 8 (1985): 245.

69. A. Schott, "An Early Account of Blood Pressure Measurement by Joseph Struthius (1510–1568)," *Medical History* 21 (1977): 305.

70. Bedford, "The Ancient Art of Feeling the Pulse," 430.

71. Marta Braun, *Picturing Time: The Work of Etienne-Jules Marey (1830–1904)* (Chicago: University of Chicago Press, 1992), 17.

72. Kuriyama, *Expressiveness of the Body*, 70.

73. Cited in Robert Michael Brain, *The Pulse of Modernism: Physiological Aesthetics in Fin-de- Siècle Europe (Seattle: University of Washington Press*, 2016), 6.

74. Brain, *The Pulse of Modernism*, 6.

75. Étienne-Jules Marey, *La méthode graphique dans les sciences expérimentales et principalement en physiologie et en médicine* (Paris: G. Masson, 1885), 108.

76. Marey, *La méthode graphique*.

77. Braun, *Picturing Time*, 19.

78. Étienne-Jules Marey, *Physiology medical de la circulation du sang* (Paris: Delahaye, 1863), 104.

79. Marey, *La méthode graphique*, 458.

80. Étienne-Jules Marey, *Movement*, trans. Eric Pritchard (New York: D. Appleton and Co., 1895), 275.

81. Marey, *Movement*, 279.

82. Étienne-Jules Marey, *La circulation du sang à l'état physiologique et dans les maladies* (Paris: G. Masson, 1881), 710; my translation.

83. Marey, *La circulation du sang*, 712.

84. Marey, 125.

85. Brain, *The Pulse of Modernism*, 21.

86. Josh Ellenbogen, *Reasoned and Unreasoned Images: The Photography of Bertillon, Galton, and Marey* (University Park, PA: Pennsylvania State University Press, 2012), 52.

87. Charlotte Stokes, "The Scientific Methods of Max Ernst: His Use of Scientific Subjects from *La Nature*," *The Arts Bulletin* 57, no. 3 (1980): 453; Brain, *The Pulse of Modernism*, 96.

88. Brain, *The Pulse of Modernism*, 96.

89. Charlotte Stokes, "The Scientific Methods of Max Ernst: His Use of Scientific Subjects from *La Nature*," *The Arts Bulletin*, LXII, no. 3 (September 1980): 453.

90. Sandra W. Moss, "The Sphygmograph in America: Writing the Pulse." *American Journal of Cardiology* 97, 4 (2006): 580–587.

91. Willem Einthoven: As part of an ongoing series of Nobel Prize winners that have contributed to cardiovascular medicine, Mark Nicholls focuses on the work of Willem Einthoven who was awarded the 1924 Nobel Prize in Physiology or Medicine "for his discovery of the mechanism of the electrocardiogram," in *European Heart Journal* 40, no. 37 (October 1, 2019): 3075–3078.

92. The illustration to the article significantly does not depict any women in the audience although there were a number present, and it did awkwardly compare the steadiness of the heartbeats of the dog and the women: "Several ladies had their heart-beats thrown on the screen also, each dipping a hand in either pot of solution. In every case their hearts were steadier than Jimmy's." "A demonstration to the Royal Society by Waller's bulldog Jimmy" *The Illustrated London News*, May 22, 1909.

93. Douglas Kahn, *Earth Sound Earth Signal: Energies and Earth Magnitude in the Arts* (Berkeley: University of California Press, 2013), 205.

94. According to the newspaper, the demonstration of the electrocardiogram was not painful because, like the humans who used that electrocardiograph, the dog only felt the sensation of salt water on his feet. However, the importance of this event lies in the fact that people began to discuss animal rights and reject their use in scientific tests.

95. P. E. Baldry, *The Battle Against Heart Disease* (Cambridge: Cambridge University Press, 1971), 75.

96. Barbara Maria Stafford, *Artful Science: Enlightenment Science and the Eclipse of Visual Education* (Cambridge, MA: MIT Press, 1994), 55.

97. See Roderick W. Home, "Fluids and Forces in Eighteenth-Century Electricity," *Endeavour* 26 (2002): 55; Paola Bertucci, "Sparks in the Dark: The Attraction of Electricity in the Eighteenth Century," *Endeavour* 31 (2007): 88; Paolo Brenni, "Jean Antoine Nollet and Physics Instruments," in *The Art of Teaching Physics: The Eighteenth-Century Demonstration Apparatus of Jean Antoine Nollet*, ed. Lewis Pyenson and Jean-François Gauvin (Quebec, Les éditions du Septentrion, 2002), 18.

98. Brenni, "Jean Antoine Nollet and Physics Instruments," 18.

99. Erkki Huhtamo, "Twin-Touch-Test-Redux: Media Archaeological Approach to Art, Interactivity and Tactility," in *Media Art Histories*, eds. Oliver Grau (Cambridge, MA; London: MIT Press, 2007), 79.

100. W. Bruce Fye, "The Origin of the Heart Beat: A Tale of Frogs, Jellyfish, and Turtles," *Circulation* 76 (1987): 493.

101. Guiliano Pancaldi, *Volta: Science and Culture in the Age of Enlightenment* (Princeton, NJ: Princeton University Press, 2003), 180.

102. Benjamin E. Jin, Heike Wulff, Jonathan Widdicombe, Jie Zheng, Donald Bers, and Jose L. Puglisi, "A Simple Device to Illustrate the Einthoven Triangle," *Advances in Physiology Education* 36, 4 (2012): 319–324.

103. J. E. Johnson, "The Nobel Prize in Physiology or Medicine 1924—Presentation Speech," Nobelprize.org, Nobel Media AB 2014, accessed July 25, 2017, http://nobelprize.org/nobel_prizes/medicine/laureates/1924/press.html

104. Johnson, "Physiology or Medicine 1924—Presentation Speech."

105. F. T. Marinetti, "The Founding and Manifesto of Futurism," in *Futurism: An Anthology*, eds. Lawrence Rainey, Christine Poggi, and Laura Wittman (New Haven, CT: Yale University Press, 2009), 49.

106. Mikhail Bulgakov, *Notes on a Cuff and Other Stories* (Richmond: London House, 2014), iv.

107. F. T. Marinetti "Multiplied Man and the Reign of the Machine" in *Futurism: An Anthology*, eds. Lawrence Rainey, Christine Poggi, and Laura Wittman (London: Yale University Press, 2009), 91.

108. F. T. Marinetti, "The Founding," 425.

109. Benedetta Cappa, "Futurist Sensibility" in in *Futurism: An Anthology*, eds. Lawrence Rainey, Christine Poggi, and Laura Wittman (London: Yale University Press, 2009), 479.

110. Michael Cowan "The Heart Machine: 'Rhythm' and Body in Weimar Film and Fritz Lang's Metropolis," *Modernism/Modernity* 14, no. 2 (2007): 225–248, https://doi.org/10.1353/mod.2007.0030.

111. George E. Burch, Nicholas DePasquale, *A History of Electrocardiography* (San Francisco: Norman, 1990).

CHAPTER 2

1. Robert Lea "Mysterious Cosmic 'Heartbeat' Detected Billions of Light-Years from Earth, *Space*, July 13, 2022, https://space.com/fast-radio-burst-cosmic-heartbeat-chime.
2. Severine Pierron, "Voyage dans le cosmos sonore d' Éliane Radigue," Centre Pompidou, November 10, 2020. https://centrepompidou.fr/fr/magazine/article/voyage-dans-le-cosmos-sonore-deliane-radigue.
3. NASA, "NASA Measuring the Pulsating Aurora," NASA, October 7, 2015, https://nasa.gov/feature/goddard/nasa-measuring-the-pulsating-aurora; Andrew Fazekas, "Pulsating Aurorae Revealed," *National Geographic*, October 3, 2010, https://nationalgeographic.com/science/article/101001-pulsating-northern-lights-aurorae-space-science.
4. Terry Ward, "There Is Little Research to Back Reports of Northern Light Sounds. So What People Are Hearing?" CNN, April 1, 2021, https://journalnow.com/news/science/there-is-little-research-to-back-reports-of-northern-lights-sounds-so-what-are-people/article_dac64c22-ec85-5f0b-9ee7-a75c98d8ef83.html?mode=nowapp.
5. Thomas Alonzo Benham, producer, "Voices of the Satellites," designer Ronald Clyne, Smithsonian Folkways Recordings.
6. Amy Shira Teitel, "Laika's Lasting Gift to American Spaceflight," *Discover Magazine*, November 3, 2017, https://discovermagazine.com/the-sciences/laikas-lasting-gift-to-american-spaceflight.
7. Teitel, "Laika's Lasting Gift."
8. Colin Burgess and Chris Dubbs, *Animals in Space: From Research Rockets to the Space Shuttle* (Chichester: Springer, 2007), 150.
9. Alice George, "The Sad, Sad Story of Laika, the Space Dog, and Her One-Way Trip into Orbit," *Smithsonian Magazine*, April 11, 2018, https://smithsonianmag.com/smithsonian-institution/sad-story-laika-space-dog-and-her-one-way-trip-orbit-1-180968728/.
10. Harry Schwartz, "Pravda Says Flight in Space Affected Laika's Heartbeat; Pravda Discloses Laika Pulse Data," *The New York Times*, May 2, 1958.
11. Larry Holzwarth, "10 Secrets of Nazi Scientists Used as Pawns in the Cold War Arms Race," *History Collection*, December 13, 2017.
12. Studies of the effects of deep space radiation and weightlessness in the cardiovascular started during the Cold War and continue until today as they still are a big concern because they can cause long-lasting damage in the cardiovascular system of astronauts. Nicola Davis, "Apollo Deep Space Astronauts Five Times

More Likely to Die from Heart Disease," *The Guardian*, 28 July, 2016, https://theguardian.com/science/2016/jul/28/apollo-deep-space-astronauts-five-times-more-likely-to-die-from-heart-disease.

13. Steven H. Platts, Michael B. Stenger, Tiffany R. Phillips, Angela K. Brown, Natalia M. Arzeno, Benjamin Levine, and Richard Summers, "Evidence Based Review: Risk of Cardiac Rhythm Problems during Space Flight," NTRS—NASA Technical Reports Server, January 1, 2009, https://ntrs.nasa.gov/citations/20100017571.

14. Al Tompkins, "War, Assassinations, Then Hope: What the Live Broadcast of the Moon Landing Meant to America," *Pyonter*, July 18, 2019, https://poynter.org/reporting-editing/2019/war-assassinations-then-hope-what-the-live-broadcast-of-the-moon-landing-meant-to-america/.

15. Cited in Marina Koren, "The Moment That Made Neil Armstrong's Heart Rate Spike," *The Atlantic*, July 15, 2019, https://theatlantic.com/science/archive/2019/07/apollo-11-moon-heart-rates/593971/.

16. W. David Woods, Kenneth D. MacTaggert, and Frank O'Brien, "Apollo Flight Journal," NASA, last modified March 2, 2021, https://history.nasa.gov/afj/ap11fj/06day2-tv.html.

17. Ashley Strickland, "NASA's 'Shout' Restores Communication with Voyager 2," CNN, August 4, 2023, https://edition.cnn.com/2023/08/04/world/nasa-voyager-2-communications-restored-scn/.

18. "Voyager, the Love Story," NASA, April 28, 2011, https://science.nasa.gov/science-news/science-at-nasa/2011/28apr_voyager2.

19. Ann Druyan, "Dear Voyagers: How Your Billion-Year Journey Carries True Love," *National Geographic*, July 10, 2019, https://nationalgeographic.com/science/article/dear-voyager-how-your-billion-year-journey-carries-true-love.

20. This is also in Yoda's words: "Age more than a count of heartbeats is."

21. Calla Cofield, "Little Green Men? Pulsars Presented a Mystery 50 Years Ago," *Space*, November 28, 2017, https://space.com/38916-pulsar-discovery-little-green-men.html.

22. Liu Jia, "No Aliens Found Yet, but 'Heartbeats' in Universe Heard," *Chinese Academy of Science*, October 17, 2017, http://english.cas.cn/newsroom/archive/news_archive/nu2017/201710/t20171017_183981.shtml.

23. The image was sourced from the Cambridge Encyclopedia of Astronomy (1977). "What Does the Cover of Joy Division's *Unknown Pleasure* Mean?" *Radio X*, November 15, 2023, https://radiox.co.uk/artists/joy-division/cover-joy-division-unknown-pleasures-meaning.

24. Samantha Mathewson, "Mysterious Gamma-Ray 'Heartbeat' Detected from Cosmic Gas Cloud," *Space*, August 19, 2020, https://space.com/mysterious-gamma-ray-heartbeat-gas-cloud.html; Ashley Strickland, "The 'Beating Hearts' of These Pulsating Stars Create Music to Astronomers' Ears," CNN, May 15, 2020, https://edition.cnn.com/2020/05/15/world/pulsating-stars-delta-scuti-scn-trnd/index.html.

25. Abdullah Alabdulgader, *Rollin McCraty, Michael Atkinson, York Dobyns, Alfonsas Vainoras, Minvydas Ragulskis, and Viktor Stoic*, "Long-Term Study of Heart Rate Variability Responses to Changes in the Solar and Geomagnetic Environment," *Scientific Reports* 8, 2663 (2018): 1.

26. Josh Sims, "Why Astronauts Are Printing Organs in Space?," *BBC*, June 2, 2021, https://bbc.com/future/article/20210601-how-transplant-organs-might-be-printed-in-outer-space.

27. Elizabeth Alvarado, "Song of Science: Musical Worlding in the American Space Age" (PhD diss., University of Chicago, 2008), 29.

28. Taylor Crumpton, "Afrofuturism Has Always Looked Forward," *Architectural Digest*, August 24, 2020.

29. Jason Heller, "Space Is the Place: The Science Fiction Pulse of '80s Electro Music," *Clarkesworld*, September 2016, 120, https://clarkesworldmagazine.com/heller_09_16/.

30. Sun Ra, "Sun Ra: The Immeasurable Equation" (Wartaweil: Waitawhile, 2005), 255.

31. Throughout his life, Milford Graves consistently worked with heartbeats. Graves created healing music melodies from the heart that, during the last years of his life, he combined with digital illustrations and medical models in sculptural pieces. A recent overview on his work can be found in the exhibition catalogue: Milford Graves, Mark Christman, ed., *Milford Graves: A Mind-Body Deal* (Los Angeles: Inventory Press, 2023).

32. Corey Kilgannon, "Finding Healing Music in the Heart," *The New York Times*, November 9, 2004.

33. Trevor Pinch, "Space Is the Place: The Electronic Sounds of Inner and Outer Space," *Journal of Sonic Studies* 8 (2014), https://researchcatalogue.net/view/108499/108500/2701/2050.

34. Heller, "Space Is the Place."

35. Michelle Kuo, "1000 Words: Kraftwerk" *Artforum*, 50, no. 9, May 2012.

36. Pauline Oliveros, *Deep Listening: A Composer's Sound Practice* (Lincoln: iUniverse, 2005), 3.

37. Oliveros, *Deep Listening*, 3.

38. Pauline Oliveros, *Sonic Meditations* (Baltimore: Smith Publications, 1971), 13.

39. Oliveros, *Sonic Meditations*, 13.

40. Larry Austin, Douglas Kahn, *Source: Music of the Avant-Garde, 1966–1973* (Berkeley: University of California Press, 2011), 244.

41. Austin and Kahn, *Source*, 245.

42. "Meditation offers significant heart benefits," Harvard Health Publishing, August 1, 2013, https://health.harvard.edu/heart-health/meditation-offers-significant-heart-benefits.

43. Pauline Oliveros, "Improvisation in the Sonosphere," *Contemporary Music Review* 25, no. 5/6, October/December (2006): 481.

44. Douglas Kahn, *Earth Sound Earth Signal: Energies and Earth Magnitude in the Arts* (Berkeley: University of California, 2013), 174.

45. Pauline Oliveros, *Sounding the Margins, Collected Writings 1992–2009* (Kingston, NY: Deep Listening Publications, 2010), 53

46. Oliveros, *Sounding the Margins*, 49.

47. Oliveros, 49.

48. Cited in Kahn, *Earth Sound Earth Signal*, 180.

49. Pauline Oliveros, "Echoes of the Moon: Notes," http://kunstradio.at/VR_TON /texte/4.html.

50. Oliveros, "Echoes of the Moon."

51. Jacques Derrida, "The Voice That Keeps Silence" in, *The Sound Studies Reader, ed.* Jonathan Sterne (New York: Routledge, 2012), 498.

52. Derrida, 498.

53. See Andreas Broeckmann, *Machine Art in the Twentieth Century* (Cambridge, MA: MIT Press, 2016), 168.

54. Tomás Saraceno's website: https://studiotomassaraceno.org/on-air-live-with/.

55. Ai Yan, "Guqin: A musical instrument that links the past with present," *CGTN*, January 4, 2018, https://news.cgtn.com/news/34556a4e35637a6333566d54/index .html.

56. Sheila Melvin, "An Ancient Chinese Instrument Is Making a Comeback," *The New York Times*, April 11, 2012. https://nytimes.com/2012/04/12/arts/12iht-guqin12 .html.

57. Ian Brennan, "Alvin Lucier: Do Something Original" *Tape Op* no. 130, March/ April 2019, https://tapeop.com/interviews/130/alvin-lucier/.

58. Richard Clar, "Giant Step: The Heartbeat and Words of Neil Armstrong," *Acta Astronautica* 186 (September 2021), 544.

59. Éliane Radigue, *Eliane Radigue* (New York: The Kitchen, 1973), http://archive .thekitchen.org/?p=7914.

60. *Omnis cellula e cellula* is a book written by German physician and biologist Rudolf Virchow, in 1855. In this text, he explains his important cell theory that rejects that organisms arise by spontaneous generation but that "all cells come from cells."

61. Éliane Radigue and Julia Eckhardt, *Éliane Radigue: Intermediary Spaces = Espaces Intermédiaires* (Brussels: Q-02 Umland, 2019), 118.

62. Éliane Radigue, email to the author, September 20, 2019.

63. Éliane Radigue, email to the author, September 20, 2019.

64. Pascal Wyse, "Interview: Eliane Radigue's Brave New Worlds," *The Guardian*, June 16, 2011.

65. Julian Cowley, "Eliane Radigue: Death Becomes Her," *The Wire* 180, February 1999. http://lovely.com/press/articles/review.radigue.wire.html.

66. Séverine Pierron, "Voyage dans le cosmos sonore d'Éliane Radigue" Centre Pompidou, November 10, 2020. https://centrepompidou.fr/en/magazine/article/voyage-dans-le-cosmos-sonore-deliane-radigue.

67. Éliane Radigue, "The Mysterious Power of the Infinitesimal," *Leonardo Music Journal* 19 (2009):

68. Geeta Dayal, "An Incandescent Force-Field: The Electronic Composer's Chry-Ptus Is Reissued," *4Columns*, October 4, 2019, http://4columns.org/dayal-geeta/eliane-radigue.

69. This happened after a concert, when some Kunga Rinpoche apprentices approached her and made her recognize the meditative quality of her music. Radigue and Eckhardt, *Intermediary Spaces*, 120.

70. Nasa Science Editorial Team, "Data Sonifications: Black Holes with a Remix," June 14, 2022, https://science.nasa.gov/learners/highlights/data-sonifications.

71. Thom Holmes, *Electronic and Experimental Music: Technology, Music, and Culture* (London: Taylor & Francis Group, 2012), 184.

72. Radigue and Eckhardt, *Intermediary Spaces*, 32.

73. Tom Johnson, "Oozing Out of the Wall," *The Village Voice*, March 29, 1973, http://tvonm.editions75.com/articles/1973/minimal-material-eliane-radigue.html.

74. Emmanuel Holterbach, "Un Portrait d'Éliane Radigue," *Le Cube*, February 7, 2014.

CHAPTER 3

1. Sabine Oelze, "Radical artist Heinz Mack turns 90," *DW*, March 3, 2021, https://dw.com/en/radical-artist-heinz-mack-turns-90/a-56807062.

2. Heinz Mack, "Germany Was Like a Vacuum—Heinz Mack on Zero," *Phaidon*, February 10, 2015, https://phaidon.com/agenda/art/articles/2015/february/10/germany-was-like-a-vacuum-heinz-mack-on-zero/.

3. Barbara Könches, *Zero Infinity Open Correspondence*, Yayoi Kusama Museum, May 29, 2020, https://yayoikusamamuseum.jp/assets/images/pdf/ZERO%20IS%20INFINITY_Open%20Correspondence_Part%20One_en.pdf.

4. Henri Neuendorf, "8 Quotes from Heinz Mack on His 85th Birthday," *Artnet*, March 8, 2016, https://news.artnet.com/art-world/heinz-mack-85-birthday-443855.

5. Edouard Derom, "The Notion of Dynamic Structure in Heinz Mack's New Concept of Painting," in *Heinz Mack. Zero Painting. Catalogue Raisonné* (Munich: Hirmer Verlag, 2017), 14.

6. Heinz Mack, *Zero* (Cambridge, MA: MIT Press, 1973), 14.

7. Mack, *Zero*, 14.

8. Mack, 40.

9. Duchamp called this "a static representation of movement." Peter Brooke, Andrew Thacker, eds., *Geographies of Modernism* (New York: Routledge, 2005), 102.
10. Mack, *Zero*, 15.
11. Gaston Bachelard, *The Dialectics of Duration* (London: Rowman & Littlefield International, 2016), 72.
12. Bachelard, *The Dialectics of Duration*, 132.
13. Henri Lefebvre, *Rhythmanalysis* (New York: Continuum, 2004), 15.
14. Bachelard, *The Dialectics of Duration*, 72.
15. Heinz Mack's website: http://mack-kunst.com/en/Graphics.htm.
16. Alan Moore, "Jean Dupuy" *Artforum* 13, no. 2 (October 1974): 73.
17. One of the first fruitful collaborations was between artist Jean Tinguely and Billy Klüver, an engineer at Bell Telephone Laboratories. Together they created *Homage to New York* (1960), a self-destroying machine that was installed in the garden of the Museum of Modern Art in New York. The Museum of Modern Art, "Homage to New York: A Self-Constructing and Self-Destroying Work of Art Conceived and Built by Jean Tinguely," press release, March 18, 1960, https://moma.org/calendar/exhibitions/3369.
18. Jean Prather, "Heart Beat Art Fascinates Onlooker," *The Republic*, April 28, 1970, 7.
19. John Cage's *Variations VII* (1966) played with sounds coming from a variety of sources present in the exhibition space, such as telephone lines, microphones and the brain waves of his collaborators on stage. For *Grass Field* (1966), artist Alex Hays was equipped with a microphone attached to his throat amplifying physiological phenomena, such as the swallowing of saliva and lung activity, although his cardiac rhythm was not heard. Catherine Morris, *9 Evenings Reconsidered: Art, Theatre, and Engineering, 1966* (Cambridge, MA: MIT Press, 2006), 9.
20. Translated from French by the author. Charles Dreyfus, "Jean Dupuy: Entretien avec Charles Dreyfus," *Érudit* 109 (2011): 19.
21. Alan Moore, "Jean Dupuy," *Artforum* 13, no. 2 (October 1974): 73.
22. The Museum of Modern Art, "The Machine as Seen at the End of the Mechanical Age," press release, November 27, 1968–February 9, 1969, https://www.moma.org/calendar/exhibitions/2776.
23. Kahn, "Let Me Hear My Body Talk," 236.
24. See Arjun Berera, "Space Dust Collisions as a Planetary Escape Mechanism" *Astrobiology* 17, no. 12 (2017): 1274–1282.
25. Charles Deyfrus, "Jean Dupuy: Entretien avec Charles Dreyfus" *Inter*, 109 (2011): 18.
26. Michael Marder, *Dust* (London; New York: Bloomsbury, 2016), 21.
27. Similarly, Turrell recalls that he "found himself staring at light shining through the air illuminating particles of dust and finding the beam more interesting that the image of the screen." Kahn, *Earth Sound Earth Signal*, 214.
28. Michael Marder, *Dust*, 21.

29. Dreyfus, "Jean Dupuy," 19.

30. Dreyfus, 18.

31. Translated from French by the author. Charles Dreyfus, "Interview Petitjean/Dupuy," *Art et nature* 65 (1996): 56.

32. This experimental group of dancers, choreographers, musicians, composers and visual artists was active from 1962–1964 and revolutionized dance aesthetics. They considered that everything from films to paintings could be seen as a dance.

33. Program for *The Visual Energy of Sound* by Jean Dupuy, Judson Church Theater, New York, November 1968.

34. *Chorus for Six Hearts* was presented alongside *Chambers* (1968) by Alvin Lucier, which required performers to create resonant environments using objects. Theodore Strongin, "Sound at Concert Makes Dust Jump," *The New York Times*, January 16, 1969.

35. Theodore Strongin, "Sound at Concert Makes Dust Jump."

36. Mark Boyle, "Journey to the Surface of the Earth/Mark Boyle's Atlas and Manual," https://boylefamily.co.uk/boyle/texts/atlas_3.html.

37. Patrick Elliott, Bill Hare, and Andrew Wilson, *Boyle Family* (Edinburgh: National Galleries of Scotland, 2003).

38. Mark Boyle, *Beyond Image: Boyle Family* (London: Arts Council for Great Britain, 1986), 50.

39. Andrew Wilson, "Towards an Index of Everything. The Events of Mark Boyle and Joan Hills 1963–71," in Elliott, Hare, and Wilson, *Boyle Family*, 50.

40. Bill Morgan, *The Beat Generation in New York: A Walking Tour of Jack Kerouac's City* (San Francisco: City Lights Books, 1997), 58.

41. Boyle, "Journey to the Surface of the Earth."

42. Boyle, "Journey to the Surface of the Earth."

43. J. L. Locher, "Mark Boyle's Journey to the Surface of Earth," https://boylefamily.co.uk/boyle/texts/journey.html.

44. Andrew Wilson, "Everything: A View on a Developing Counterculture in the Mid 1960s in London," in *Blast to Freeze. British Art in the 20th Century*, eds. Gijs Tujl and Henry Meyric-Hughes (Wolfsburg: Kunstmuseum Wolfsburg, 2002), 215.

45. Martin Esslin, *Antonin Artaud* (London: Penguin Books, 1977), 87.

46. The participation of Mark Boyle in Schneemann's *Meat Joy* is registered in the Boyle's family website, https://boylefamily.co.uk/boyle/info/index.html.

47. Maurice Merleau-Ponty, *Phenomenology of Perception* (New York: Routledge, 1962), 94.

48. Locher, "Mark Boyle's Journey to the Surface of Earth."

49. Locher, "Mark Boyle's Journey to the Surface of Earth."

50. Locher, "Mark Boyle's Journey to the Surface of Earth."

51. Locher, "Mark Boyle's Journey to the Surface of Earth."
52. Locher, "Mark Boyle's Journey to the Surface of Earth."
53. Ecole Polytechnique Fédérale de Lausanne. "Your Brain Suppresses Perception Related of Heartbeat, for Our Own Good," EPFL, May 4, 2016, https://actu.epfl.ch/news/your-brain-suppresses-perception-of-heartbeat-for-/.
54. One of the Boyle Family's exhibition at Luhring Augustine Gallery bore that sentence as its title in 2021.
55. Margarita Schultz, "Interview with Diana Domingues," *Heterogénesis*, July 24, 1998.
56. This medical instrument became popular in the health system in the 1960s because it provided a real time visualization of heart rates.
57. Diana Domingues, *Pulsação Ou Onda Ou Vermelho, Não Verde* (São Paolo: Fundação Armando Alvares Penteado, 1986), 45.
58. Interview with Diana Domingues conducted by the author, March 1, 2022.
59. N. Katherine Hayles, *How We Became Posthuman* (Chicago: Chicago University Press, 1992), 56.
60. Interview with Diana Domingues conducted by the author, March 1, 2022.
61. Diana Domingues, "Interactivity and Ritual: Body Dialogues with Artificial Systems" (SIGGRAPH, Los Angeles, 1999).
62. Domingues, "Interactivity and Ritual."
63. Domingues, "Interactivity and Ritual."
64. René Guenon, "The Heart and the Cave," *Studies in Comparative Religion* 5, no. 1 (1971): 24.
65. Valerie de Marinis, "With Dance and Drum: A Psychocultural Investigation of the Ritual Meaning-Making System of an Afro-Brazilian," *Scripta Instituti Donneriani Aboensis 16* (1996): 71.
66. Michael Heckenberger, "Xinguano, Ancestors, and Others," in Time and Memory in Indigenous Amazonia, ed. Carlos Fausto and Michael Heckenberger (Miami: University Press Florida, 2013), 294.
67. Stewart Ain, "A New Study Has Found That Heartbeats Actually Harmonize during This Religious Ritual," *Forward*, March 24, 2021.
68. Bruno Gringas, Gerard Pohler, and W. Tecumseh Fitch "Exploring Shamanic Journeying: Repetitive Drumming with Shamanic Instructions Induces Specific Subjective Experiences but No Larger Cortisol Decrease than Instrumental Meditation Music," *PLOS ONE* 9, no. 7 (2014): e102103.
69. Domingues, "Interactivity and Ritual."
70. Roy Ascott, "Weaving the Shamantic Web: Art in the Bio-telematic Domain" (Unpublished manuscript, 1997), PDF.
71. Char Davies, "OSMOSE: Notes on Being in Immersive Virtual Space," *Digital Creativity* 9, no. 2 (1998): 67.

72. Diana Domingues, "Ouroboric Perception and the Effects of Enactive Affective Systems to the Naturalization of Technologies," *Media-N* 12, no. 1 (2016): 35.

73. Domingues, "Ouroboric Perception," 30.

74. This is a term coined by the artist, with which she refers to the coexistence of physical and cyber worlds.

75. Ascott, "Weaving the Shamantic Web."

CHAPTER 4

1. Mark Porter, "How Your Life Expectancy Is Affected by Your Heart Rate," *The Times*, February 11, 2020.

2. Tahnée Engelen, M. Solcà, C. Tallon-Baudry, "Interoceptive Rhythms in the Brain," *Nature Neuroscience* 26, (2023): 1670.

3. Allan Kaprow from his manuscript lecture for "Time Pieces," quoted in Glen Phillips, "Time Pieces," in *Allan Kaprow: Art as Life*, ed. Eva Meyer-Hermann, Andrew Perchuk, and Stephanie Rosenthal (Los Angeles: Getty Research Institute, 2008), 36.

4. Phillips, "Time Pieces," 36.

5. Interview with Julieta Aranda conducted by the author, April 9, 2021.

6. This technology is a portable version of an EKG. It is more likely to detect an arrythmia because of its longer recording time.

7. Augustine, *The Confessions* (Indianapolis: Hackett Publishing Company, 2006), 242.

8. Henri Bergson, *Matter and Memory*, translated by Nancy Margaret Paul and W. Scott Palmer (London: George Allen and Unwin, 1911), 55.

9. Emmanuel Lévinas, *Otherwise Than Being, or Beyond Essence* (Pittsburgh: Duquesne University Press, 1998), 181.

10. "Study Finds Perception of Time Linked to Heartbeat," *Science Daily*, March 7, 2003, https://sciencedaily.com/releases/2023/03/230307114434.htm.

11. Jürgen Portschy, "Times of Power, Knowledge and Critique in the Work of Foucault," *Time & Society* 29, no. 2 (2020): 393.

12. Sarah Sharma, "The Biopolitical Economy of Time," *Journal of Communication Inquiry* 35, no. 4 (2011): 442.

13. Velimir Khlebnikov, *Collected Works of Velimir Khlebnikov*, vol. I (Cambridge, MA: Harvard University Press, 1987), 357.

14. Rasa Tsuda, "Heartbeat Is the Beginning: Interview with Heartbeat Sasaki," *Azito*, 2013, http://azito-art.com/topics/heartbeat-is-the-beginning-interview-with-heartbeat-sasaki/.

15. Tsuda, "Heartbeat Is the Beginning."

16. Sophie Duvernoy, "The Beating of My Heart," *LA Weekly*, July 8–14, 2011.

17. See Broeckmann, *Machine Art in the Twentieth Century*, 210.
18. Sasaki's official website: http://heartbeatdrawing.com/.
19. Tsuda, "Heartbeat Is the Beginning."
20. Sasaki, email correspondence between the artist and the author, July 22, 2020.
21. Mami Kataoka, "Looking at the World as an Accumulation of Pulsations," in *Project N 04*, ed. Tokyo Opera City Art Gallery, 2000, http://heartbeatdrawing.com/text/text-2.html.
22. Blood circulation has a circadian clock because blood vessels work in a feedback loop of twenty-four hours keeping blood pressure steady during the day and fall during night. For further information on the circadian clock and its vascular implications read Daniel Ruic, "Time Is of the Essence: Vascular Implications," *Circulation* 120, no. 17 (2009): 1714–1721.
23. Tsuda, "Heartbeat Is the Beginning."
24. Sasaki, email correspondence between the artist and the author, July 21, 2020.
25. Sasaki, email correspondence between the artist and the author, July 21, 2020.

CHAPTER 5

1. Ayesha Nathoo, *Hearts Exposed. Transplants and the Media in 1960s Britain* (London: Palgrave Macmillan, 2009), 63.
2. Brian O'Doherty, "Strolling with the Zeitgeist," *Frieze* 153 (March 2013): 139.
3. Lucy Lippard, *Six Years: The Dematerialization of the Art Object from 1966 to 1972* (Berkeley: University of California Press, 1973).
4. Ayesha Nathoo, "The Transplanted Heart: Surgery in the 1960s," in *The Heart*, ed. James Peto (New Haven, CT: Yale University Press, 2007), 156.
5. This was responding to the fact that the second heart transplant, performed in January 2, 1968, was from Clive Haupt, a coloured man, whose heart was given to Phillip Blaiberg, a white South African. This transcended the discriminatory policies of racial separation that were in place in South Africa at the time. Ali Mazrui, "The Poetics of a Transplanted Heart," *Transition* 35 (1968): 51.
6. Herb Boyd, "Dr. Myra Adele Logan, First Woman to Perform Open-Heart Surgery," *Amsterdam News*, November 14, 2019, https://amsterdamnews.com/news/2019/11/14/dr-myra-adele-logan-first-woman-perform-open-heart/.
7. Mazrui, "The Poetics," 58.
8. Mazrui, "The Poetics," 58.
9. Brian O'Doherty, "The Politics and Esthetics of Heart Transplants," *Art International* 15 (1971): 27.
10. Brian O'Doherty, "Taking Duchamp's Portrait," lecture, Fourth Edgar P. Richardson Symposium on Portraiture, Donald W. Reynolds Center for American Art and Portraiture, Washington DC, March 27, 2009. The symposium was

presented in conjunction with the exhibition *Inventing Marcel Duchamp: The Dynamics of Portraiture*, at the Smithsonian's National Portrait Gallery, https://youtube.com/watch?v=jBAa9646yOw.

11. O'Doherty, "Taking Duchamp's Portrait."
12. O'Doherty, "Taking Duchamp's Portrait."
13. Jacques Derrida, *Speech and Phenomena: And Other Essays on Husserl's Theory of Signs*, trans. David Allison (Evanston, IL: Northwestern University Press, 1973), 156.
14. Jean-Luc Nancy, *Le regard du portrait* (Paris: Éditions Galilée, 2000), 12.
15. Nancy, *Le regard du portrait*, 29.
16. Margaret Iversen, *Photography, Trace and Trauma* (Chicago: University of Chicago Press, 2017), 75.
17. O'Doherty, "Taking Duchamp's Portrait."
18. O'Doherty, "Taking Duchamp's Portrait."
19. O'Doherty, "Taking Duchamp's Portrait."
20. Thomas Hess, "J'accuse Marcel Duchamp," *Art News* 43, no. 10 (1965): 44–45, 52–54.
21. O'Doherty, "Taking Duchamp's Portrait."
22. Linda Dalrymple Henderson, *Duchamp in Context: Science and Technology in the Large Glass and Related Works* (Princeton, NJ: Princeton University Press, 1998), 35.
23. Cited in Iversen, *Photography, Trace and Trauma*, 75.
24. Pierre Cabanne, *Dialogues with Marcel Duchamp* (New York: Viking Press, 1971), 68.
25. Rosalind Krauss, *Formless: A User's Guide* (Cambridge, MA: MIT Press, 1997), 133.
26. Brian O'Doherty, "Taking Duchamp's Portrait."
27. Catherine Richards, *Catherine Richards: Coeurs électrisés* (Ottawa: Musée des beaux-arts du Canada, 1997), n.p.
28. Catherine Richards, "Artist Statement: Charged Hearts," http://catherinerichards.ca/artwork2/Hearts_statement.html.
29. Richards, "Artist Statement."
30. Donna Haraway, *Simians, Cyborgs and Women: The Reinvention of Nature* (New York: Routledge, 1991), 149–150.
31. Catherine Richards, *Excitable Tissues* (Ottawa: The Ottawa Art Gallery, 2003), 100.
32. Kim Sawshuck, "Catherine Richards," *Parachute* 89 (1998): 49–50.
33. Dot Tuer, *Mining the Media Archive: Essays on Art, Technology, and Cultural Resistance* (Toronto: YYZ Books), 28.
34. See Hybrid Bodies website: http://hybridbodiesproject.com/.
35. See Oliver E. Mauthner, Enza De Luca, Jennifer M. Poole, Susan E. Abbey, Margrit Shildrick, Mena Gewarges, and Heather J. Ross, "Heart Transplants: Identity Disruption, Bodily Integrity and Interconnectedness," *Health* 19, no. 6 (2015): 578; Ana M. Palmar-Santos, Azucena Pedraz-Marcos, Juan Zarco-Colón,

Milagros Ramasco-Gutiérrez, Eva García-Perea, and Monserrat Pulido-Fuentes, "The Life and Death Construct in Heart Transplant Patients," *European Journal of Cardiovascular Nursing* 18, no. 1 (2019): 48.

36. Jean-Luc Nancy, "L'Intrus," *CR: The New Centennial Review* 2, no. 3 (2002): 10.
37. Nancy, "L'Intrus," 7.
38. Nancy, 7.
39. Catherine Richards, "Artist Statement: L'intrus," http://catherinerichards.ca/artwork2/LIntrus_statement.html.
40. Marshall McLuhan, *Understanding Media: The Extensions of Man* (Cambridge, MA: MIT Press, 1994), 314.

    cited in Erkki Huhtamo, "Twin-Touch-Test-Redux: Media Archeological Approach to Art, Interactivity and Tactility," in *Media Art Histories* (Cambridge, MA: The MIT Press, 2007), 71.
41. Annerose Böhrer and Larissa Pfaller, "The Tell-Tale Heart: Multiple Ontologies of the First Human Donor Heart," in *The Gender of Things, ed.* Maria Rentetzi (New York: Routledge, 2024), 91.
42. Palmar-Santos et al., "The Life and Death Construct," 53.
43. Catherine Richards, "Artist Statement: This Isn't All the Hearts I Have," http://catherinerichards.ca/artwork2/ThisIsntAll_statement.html.
44. Catherine Richards, "Artist Statement: I Can't Let Go of Them," http://catherinerichards.ca/artwork2/ICantLetGo_statement.html.

CHAPTER 6

1. Irma van der Ploeg, "Biometrics and the Body as Information," in *Surveillance as Social Sorting: Privacy, Risk and Automated Discrimination*, ed. David Lyon (New York: Routledge, 2002): 64; David Lyon, *Biometrics, Identification and Surveillance*, *Bioethics* 22, no. 9 (2008): 499.
2. Btihaj Ajana, *Governing Through Biometrics: The Biopolitics of Identity* (London: Palgrave Macmillan, 2013), 44.
3. Ajana, *Governing Through Biometrics*, 7.
4. *Information Technology: Its New Meaning for Art* in 1970 at the Jewish Museum in New York. *Art and Technology* in 1971 at Los Angeles County Museum of Art, which didn't include any female artists.
5. Teresa Burga in interview conducted by the author, June 22, 2018.
6. Teresa Burga in interview conducted by the author, June 22, 2018.
7. Jean-Luc Nancy, *Corpus* (New York: Fordham University Press, 2008), 150.
8. Nancy, *Corpus*, 155.
9. Lucy R. Lippard, *Six Years: The Dematerialization of the Art Object from 1966 to 1972* (Los Angeles: University of California Press, 1997), viii.

10. Teresa Burga in interview conducted by the author, June 22, 2018.
11. Miguel A. López, "As Near as possible to Chance" A Dialogue with Teresa Burga" in *Teresa Burga: Estructuras de aire = structures of air, eds.* Miguel A. López and Agustin Pérez (Buenos Aires: MALBA Fundación Eduardo F. Costantini, 2015).
12. Elisa Arca, José-Carlos Mariategui, "Sistemas de información en el trabajo de Teresa Burga" *Revista Ui*, 4, no. 5 (2020): 72.
13. Teresa Burga in interview conducted by the author, June 22, 2018.
14. Teresa Burga in interview conducted by the author, June 22, 2018.
15. Lippard, *Six Years*, xv.
16. Felix Stalder and David Lyon, "Electronic Identity Cards and Social Classification," in *Surveillance as Social Sorting: Privacy, Risk and Automated Discrimination*, ed. David Lyon (Abingdon: Routledge, 2002), http://ebookcentral.proquest.com/lib/unsw/detail.action?docID=240591.
17. See Michel Foucault, *Discipline and Punish: The Birth of Prison* (New York: Vintage Books, 1977), 191; Michel Foucault, *Security, Territory, Population* (London: Palgrave, 2007), 1.
18. Roberto Esposito, *Terms of the Political: Community, Immunity, Biopolitics*, trans. Rhiannon Noel Welch (New York: Fordham University Press, 2013), 71.
19. Ajana, *Governing Through Biometrics*, 7.
20. Dorota Biczel, "Cracking Open the Systems: Media, Materiality, and Agency in Teresa Burga's *Self-Portrait. Structure. Report. 9.6.72*," *Alter/Nativas* 3 (2014), https://alternativas.osu.edu/assets/files/Issue-3/essays/biczel.pdf.
21. Teresa Burga, in Lopez and Rubio, *Estructuras de aire/Structures of Air*.
22. See Anil K. Jain, Patrick Flynn, and Arun A. Ross, eds. *Introduction to Biometrics* (Berlin: Springer, 2011), 97–139.
23. George E. Billman, "Heart Rate Variability," *Frontier in Physiology* 2 (2011): 86.
24. Haraway, *Simians, Cyborgs and Women*, 164.
25. Michael Golston, *Rhythm and Race in Modernist Poetry and Science* (New York: Columbia University Press, 2008), ii.
26. Arturo Obregoso, "Eugenesia, tests mentales y degeneración racial en el Perú," *Revista de Psicología Trujillo* 14, *no.* 2 (2012): 232; Laura Cházaro, "La fisioantropometría de la respiración en las alturas, un debate por la patria" *Ciencias* 60, 61 (2000): 40.
27. Catherine Grenier, *Mona Hatoum* (Vancouver, BC: Rennie Collection, 2010), 9.
28. Michael Archer, Catherine de Zegher, and Guy Brett, *Mona Hatoum* (London: Phaidon, 1997), 119.
29. Mona Hatoum, "The John Tusa Interviews," BBC Radio 3, September 4, 2005, https://bbc.co.uk/sounds/play/pooncz8t.
30. Bill Viola, *Bill Viola: Installations and Videotapes* (New York: The Museum of Modern Art, 1987), 48.

31. Lynda Birke, *Feminism and the Biological Body* (Edinburgh: Edinburgh University Press, 1999), 47.

32. Mark Brown, "Mona Hatoum's Internal Organs to Feature in Tate Modern Show," *The Guardian*, May 3, 2016.

33. Janine Antoni and Mona Hatoum, "Mona Hatoum," *BOMB* 54 (1998): 60.

34. Archer et al., *Mona Hatoum*, 25.

35. Emma Robertson, "Mona Hatoum," *The Talks*, October 19, 2016, https://the-talks.com/interview/mona-hatoum/

36. Bruce Curtis, "Foucault on Governmentality and Population: The Impossible Discovery," *Canadian Journal of Sociology* 27, no. 4 (2002): 515; Michel Foucault, *Discipline and Punish*.

37. Donna Haraway, *Modest_Witness@Second_Millennium. FemaleMan©_Meets_OncoMouse™* (New York: Routledge, 1997), n.p.

38. Elena Tzelepis, "Vulnerable Corporealities and Precarious Belongings in Mona Hatoum's Art," in *Vulnerability in Resistance*, eds. Judith Butler, Zeynep Gambetti, and Leticia Sabsay (Durham and London: Duke University Press, 2016), 151.

39. David Armstrong, "The Rise of Surveillance Medicine," *Sociology of Health and Illness* 17, no. 3 (1995): 393–404.

40. Armstrong, "The Rise of Surveillance Medicine," 395.

41. "Environment, Climate Change and Heart Health," *The Heart Foundation*, February 21, 2024, https://heartfoundation.org.au/healthy-living/physical-activity/climate-change-and-heart-health.

42. Kate Crawford and Jason Schultz, "AI Systems as State Actors," *Columbia Law Review* 119, no. 7 (November 2019): 1942; van der Ploeg, "Biometrics and the Body," 58.

43. Emily Tat, Deepak Bhatt, and Mark G Rabbat "Addressing Bias: Artificial Intelligence in Cardiovascular Medicine" *The Lancet Digital Health* 12, no. 2 (December 2020): E635–636.

44. David Armstrong, "Bodies of Knowledge: Foucault and the Problem of Human Anatomy," in *Sociological Theory and Medical Sociology*, ed. G. Scambler (London: Tavistock, 1987), 70 (brackets added).

45. Some of the material in this section on Lozano-Hemmer has been drawn from the following publications: Claudia Arozqueta, "Heartbeats and the Arts: A Historical Connection," *Leonardo* 51, no. 1(2018): 33–39; Claudia Arozqueta, "Polysemous Light: Light as a Contradictory Material in Rafael Lozano-Hemmer's Work," *Artnodes* 21 (2018), http://dx.doi.org/10.7238/a.voi21.3185; Claudia Arozqueta, "Connecting Realities: Rafael Lozano-Hemmer's Pulse-Based Works," *Journal of Science and Technology of the Arts* 6, no. 2 (2014): 21–28, https://doi.org/10.7559/citarj.v6i2.117; Claudia Arozqueta "Seeing the Unseen: Rafael Lozano-Hemmer's Pulse Works," *Journal of Science and Technology of the Arts* 6, no. 2 (2015).

46. Rafael Lozano-Hemmer and Rafael Lozano-Hemmer, *Some Things Happen More Often than All of the Time* (Mexico City: Turner, 2007), 25.

47. Sadie Dingfelder, "This New Hirshhorn Exhibit Will Turn Your Heart Into Art," *The Washington Post*, October 30, 2018.

48. Dingfelder, "This New Hirshhorn Exhibit."

49. Peter Sloterdijk, "The Open Clearing and Illumination. Remarks on Metaphysics, Mysticism and the Politics of Light," in *Light Art from Artificial Art: Light as a Medium in 20th and 21st Century Art*, eds. Peter Weibel and Gregor Jansen (Ostfildern: Hatje Cantz, 2006), 50.

50. Elizabeth Fortescue, "Artwriter Interviews Rafael Lozano-Hemmer," *Artwriter*, December 2011, http://artwriter.com.au/news/artwriter-interviews-rafael-lozano-hemmer.

51. All these artworks and more are part of an archive of more than 100 pieces that I started to collect in 2010 while writing the MA thesis "Seeing the Unseen: Rafael Lozano-Hemmer's works," which includes a chapter on heartbeat art works.

52. Marie-Pier Boucher and Patrick Harrop, "Interview with Rafael Lozano-Hemmer," *Inflexions* 5 (2012): 160.

53. Allison Marsh, "A Brief History of the Lie Detector," *IEEE Spectrum*, July 21, 2019, https://spectrum.ieee.org/a-brief-history-of-the-lie-detector.

54. See Claudia Arozqueta, "In the Spotlight: Searchlights, Art, Surveillance and Spectacle," LUX AETERNA, Proceedings of the 25th International Symposium on Electronic Art (ISEA), Asia Culture Center, Gwangju, South Korea, 2019, 24–28.

55. Rafael Lozano-Hemmer's official website: https://lozano-hemmer.com/bio.php.

56. James C. Ross, "Datamigrants," *Radical Philosophy* 140 (November/ December 2006): 13.

57. This is addressed in Giorgio Agamben, *State of Exception* (Chicago and London: The University of Chicago Press, 2005).

58. Ross, "Datamigrants," 11.

59. David Hambling, "The Pentagon Has a Laser That Can Identify People from a Distance—by Their Heartbeat," *MIT Technology Review*, June 29, 2019, https://technologyreview.com/2019/06/27/238884/the-pentagon-has-a-laser-that-can-identify-people-from-a-distanceby-their-heartbeat/.

60. Lozano-Hemmer, correspondence with the author, August 10, 2020.

61. Dingfelder, "This New Hirshhorn Exhibit."

62. Tessa Koumoundouros, "People's Heartbeats Synchronize When They're Captivated by the Same Story," *Science Alert*, September 15, 2021; Delphine Grynberg and Olga Pollatos, "Perceiving One's Body Shapes Empathy," *Physiology and Behavior* 140 (2015): 54–60; R. Michael Winters, Bruce N. Walker, and Grace Leslie, "Can You Hear My Heartbeat?: Hearing an Expressive Biosignal Elicits Empathy" *CHI '21: Proceedings of the 2021 CHI Conference on Human Factors in Computing Systems* no. 225 (2021): 1–11.

63. The work can be found at http://onpulse.net. Information shared by the artist through the email correspondence with the author, August 11, 2020.

64. Claire Fahy, "Runners and Cyclists Use GPS Mapping to Make Art," *The New York Times*, September 24, 2022, https://nytimes.com/2022/09/24/technology/gps-art-strava-running.html.

65. Steven A. Israel, "The Heartbeat: The Living Biometric," in *Biometrics: Theory, Methods and Applications*, ed. Evangelia Micheli-Tzanakou, Konstantinos N. Plataniotis, and N. V. Boulgouris (Hoboken, NJ: John Wiley & Sons, 2009), 429.

66. Donna Haraway, *Simians, Cyborgs and Women: The Reinvention of Nature* (New York: Routledge, 1991), 150.

67. N. Katherine Hayles, *How We Became Posthuman* (Chicago: Chicago University Press, 1992), 56.

68. Hayles, *How We Became Posthuman*, xiii.

69. Hayles, 3.

70. Irma van der Ploeg, "Biometrics and the Body as Information," 64.

71. Kate Crawford, *Atlas of AI* (London: Yale University Press, 2021), 211.

CHAPTER 7

1. See Sian E. Harding, *The Exquisite Machine* (Cambridge, MA: MIT Press, 2022), 69–70.

2. Yoko Ono, *Grapefruit* (New York: Simon & Shuster, 1970), n.p.

3. Ono, *Grapefruit*.

4. Alexandra Munroe, *Japanese Art Since 1945: Scream Against the Sky* (Harry N. Abrams: New York, 1996), 220.

5. Yoko Ono has claimed this repeatedly in interviews, texts, and on social media. Midori Yoshimoto, *Into Performance: Japanese Women Artists in New York* (New Brunswick, NJ: Rutgers University Press, 2005), 79; Yoko Ono (@yokoono), "Art is a way of survival," Twitter, December 15, 2012, 4:47 p.m., https://twitter.com/yokoono/status/280079152739123201.

6. In 1964, Ono self-published *Grapefruit* in a small edition of 500 copies in Japan. Its second edition was published by Simon & Shuster in New York, in 1970, with an introduction by John Lennon.

7. Cott, "Yoko Ono and Her Sixteen-Track Voice," *Rolling Stone*, March 18, 1971, https://rollingstone.com/music/music-news/yoko-ono-and-her-sixteen-track-voice-237782.

8. Ono, *Grapefruit*.

9. Yoko Ono (@yokoono), "Grapefruit is a hybrid of lemon and orange. Snow is hybrid of wish and lament," Twitter, December 10, 2014, 3:50 p.m., https://twitter.com/yokoono/status/542798410638557184.

10. Jillian Suarez, "The Hand-Lettered Title on Yoko Ono's Artist Book," Guggenheim, April 9, 2015, https://guggenheim.org/blogs/findings/the-hand-lettered-title-on-yoko-onos-artist-book.

11. Alexandra Munroe, "Why War? Yoko by Yoko at the Serpentine," in *Yoko Ono: To the Light, eds. Kathryn Rattee, Melissa Larner, and Rebecca Lewin* (London: Serpentine Gallery; Cologne: Koenig Books, 2012), 13.
12. Hans Ulrich Obrist, "Mix a Building and the Wind: An Interview of Yoko Ono by Hans Ulrich Obrist," *e-flux*, 2002, http://projects.e-flux.com/do_it/notes/interview/i002.html.
13. Yoshimoto, *Into Performance*, 18.
14. Obrist, "Mix a Building and the Wind."
15. Alexandra Munroe, "Spirit of YES: The Art and Life of Yoko Ono," in *Yes Yoko Ono*, eds. Alexandra Munroe, Yoko Ono, Jon Hendricks, and Bruce Altshuler (New York: Japan Society, 2000), 10–37.
16. See chapter 1.
17. Shigehisa Kuriyama, *The Expressiveness of the Body and the Divergence of Greek and Chinese Medicine* (New York: Zone Books, 1999), 135.
18. Ono, *Grapefruit*.
19. Yoko Ono, "To the Wesleyan People," in *Yoko Ono: To See the Skies*, ed. Jon Hendricks (Milan: Mazzotta, 1990).
20. Katia Lenehan, ed., *Spiritual Foundations and Chinese Culture: A Philosophical Approach* (Washington: Council for Research in Values and Philosophy, 2016), 220.
21. Kuriyama, *The Expressiveness of the Body*, 93–96.
22. Kuriyama, 99.
23. Yoko Ono quoted on MOMA's website, https://moma.org/audio/playlist/15/372.
24. Jacques Derrida, *On Touching—Jean-Luc Nancy*, trans. Christine Irizarry (Stanford, CA: Stanford University Press, 2005), 18.
25. Jean Luc-Nancy, *The Muses* (Stanford: Stanford University Press, 1993), 17.
26. Jean Luc-Nancy, *Dis-Enclosure: The Deconstruction of Christianity* (New York: Fordham University Press, 2008), 80.
27. Liz Kotz, "Post-Cagean Aesthetics and the 'Event' Score," *October* 96 (2001): 50.
28. Ono, "To the Wesleyan People," 13.
29. Obrist, "Mix a Building and the Wind."
30. Thomas Britt, "It's Me, I'm Alive: A Conversation with Yoko Ono," *Pop Matters*, February 7, 2010, https://popmatters.com/yoko-ono-interview-2496153966.html.
31. Baby heartbeats have been used politically. Anne Pollock has addressed the use of the fetal heartbeat and its political connotations in relation to recent abortion bills. Anne Pollock, "Heart Feminism," *Catalyst: Feminism, Theory, Technoscience* 1 (2015): 14.
32. Pollock, "Heart Feminism."
33. Neda Ulaby, "Yoko Ono Joins John Lennon with Credit Line for Writing 'Imagine,'" *NPR*, June 17, 2017, https://npr.org/2017/06/17/533368546/.

34. Ono, *Grapefruit.*

35. Yoshimoto, *Into Performance*, 114.

36. Billy Amendola "Yono Ono Talks Drums with Modern Drummer," *Modern Drummer*, November 2009. https://moderndrummer.com/2009/11/yoko-ono/.

37. Johnstone, *Yoko Ono (Talking)* (London: Omnibus Press, 2006), 54.

38. "Why" appears on her album *Yoko Ono/Plastic Ono Band*, released in 1970. Cott, "Yoko Ono and Her Sixteen-Track Voice."

39. Yoko Ono (@yokoono), "Our hearts beat in unison. Even when we fight with one another. We are indeed one world, one people," Twitter, June 25, 2009, 2:00 a.m., https://twitter.com/yokoono/status/2312134676.

40. Yoko Ono (@yokoono), "Listen to the heartbeat of all things on the planet. It is symphonic. The prima heartbeat is that of the Ocean," Twitter, December 21, 2016, 12:44 p.m., https://twitter.com/yokoono/status/811643397496832004.

41. "Yoko Ono on Survival, Anger and Occupying White Space," *Paper*, July 24, 2018. https://papermag.com/bell-hooks-yoko-ono-1425142876.html?rebelltitem=14#rebelltitem14.

42. This work by Linda Montano is also titled *Listening to My Heart . . . A Congenital Murmur*. After having exchanged some emails, Montano eventually remembered this work after I sent her a documentation image that I found in the book Linda Montano, *Linda Mary Montano: You Too Are a Performance Artist* (Santa Fe, NM: SITE Santa Fe, 2013). Linda Montano, email correspondence between the artist and the author, July 11, 2017.

43. Linda Montano, *Letters from Linda M. Montano* (New York: Routledge, 2005), 25.

44. Linda Montano, email correspondence between the artist and the author, July 11, 2017.

45. Jennifer Douglas, Brittany Matheson, and June Liang. (2015). "The Relationship between Eating Disorders and Sexual Trauma," *Eating and Weight Disorders* 20, no. 2, https://doi.org/10.1007/s40519-015-0195-y.

46. Elizabeth Stephens and Annie Sprinkle, "Linda M. Montano's Archive," and "Art/Life with Linda M. Montano," *Sexecology*, http://sexecology.org/research-writing/linda-montano.

47. Many scholars have noted that endurance art is a vague term and does not necessarily designate a movement. As a term endurance art has often been used interchangeably with masochist art and durational art, the latter sharing a common etymological root in the Latin word *durare*, "to last." Lara Shalson, *Performing Endurance: Art and Politics Since 1960* (Cambridge: Cambridge University, 2018), 3.

48. Karen Gonzalez Rice, *Long Suffering: American Endurance Art as Prophetic Witness* (Ann Arbor, MI: University of Michigan Press, 2016), 4.

49. Ulay considers that the brain plays a secondary role compared to that one of the heart "for the simple reason that emotions conditions the whole

chemistry and thinking." Cited in Linda Montano, *Performance Artists Talking in The Eighties* (Berkeley and Los Angeles: University of California Press, 2000), 332.

50. Linda Montano, correspondence between the artist and the author, February 4, 2020.

51. Donna Haraway, *The Companion Species Manifesto: Dog, People, and Significant Otherness* (Chicago: Prickly Paradigm Press, 2002), 12.

52. Montano, correspondence with the author, February 4, 2020.

53. Montano, correspondence between the artist and the author, July 9, 2019.

54. Montano, correspondence with the author, July 9, 2019.

55. The Order of Saint Benedict, "Prologue of the Rule of Saint Benedict," http://archive.osb.org/rb/text/rbejms1.html.

56. See Raymond H. Hull and Stacy R. Kerschen, "The Influence of Cardiovascular Health on Peripheral and Central Auditory Function in Adults: A Research Review," *American Journal of Audiology* 19, no. 1 (2010): 9–16; Harvard Health Publishing, "Takotsubo Cardiomyopathy (Broken-Heart Syndrome)," January 29, 2020, https://health.harvard.edu/heart-health/takotsubo-cardiomyopathy-broken-heart-syndrome.

57. Charles Stankievech, "From Stethoscopes to Headphones: An Acoustic Spatialization of Subjectivity," *Leonardo Music Journal* 17 (2007): 56.

58. Jennie Klein, "Feeding the Body: The Work of Barbara Smith," *PAJ: A Journal of Performance and Art* 21, no. 1 (1999): 25.

59. Pauline Oliveros, *Sonic Meditations* (Baltimore: Smith Publications, 1971), 2.

60. Pauline Oliveros, *Deep Listening: A Composer's Sound Practice* (Lincoln: iUniverse, 2005), 31

61. Oliveros, *Deep Listening*, 42

62. Oliveros, 21

63. Jennifer Fisher, "The Chakra Cycles of Linda M. Montano," in *Technologies of Intuition*, ed. Jennifer Fisher (Toronto: YYZ Books, 2006), 158.

64. Montano, *Letters from Linda M. Montano*, 242.

65. Montano, *Letters from Linda M. Montano*, 228.

66. Terence Handley Macmath, "Interview: Linda Mary Montano, Performance Artist," *Church Times*, May 13, 2015.

67. Ludwig Feuerbach, *The Essence of Christianity* (New York: Calvin Blanchard, 1855), 54.

68. Luce Irigaray, *To Be Two* (New York: Routledge, 2001), 11.

69. Violet Snow, "Saugerties Church Hosts Linda Montano Performance Piece," *HV1*, October 11, 2018, https://hudsonvalleyone.com/2018/10/11/saugerties-church-hosts-linda-montano-performance-piece.

70. Montano, correspondence with the author, February 4, 2020.

71. "Claw of Death Is 'Hand of God' at Giant Paris Art Show," *Mutual Art*, January 13, 2010, https://mutualart.com/Article/Claw-of-death-is--hand-of-God--at-giant-/9CB53E3DDF07D838.

72. Tamar Garb, "In Conversation with Christian Boltanski," in *Christian Boltanski, eds.* Didier Semin, Christian Boltanski, Tamar Garb, and Donald Burton Kuspit (London: Phaidon, 1997), 19.

73. Daniel Mendelsohn, *Christian Boltanski* (Paris: Flammarion, 2010), 142.

74. Laura Cumming, "Christian Boltanski: Personnes," *The Guardian*, January 17, 2010, accessed May 12, 2020, https://theguardian.com/artanddesign/2010/jan/17/christian-boltanski-personnnes-paris-review.

75. Mendelsohn, *Christian Boltanski*, 144.

76. Andrew Maerkle, "Memory, Too, Fails at Such Excess," *Art It*, July 23, 2010, https://art-it.asia/en/u/admin_ed_itv_e/zbh3nmp807ggve9tiqac.

77. Peter Aspden, "Christian Boltanski: 'We Give Ourselves Things to Do So That We Don't Die,'" *Financial Times*, March 30, 2018, https://ft.com/content/182368ec-3212-11e8-ac48-10c6fdc22f03.

78. Nicole Billitz, "Finitude, Existence, and Community: Letting the Individual Die," *Inquiries* 4, no. 7 (2012), http://inquiriesjournal.com/a?id=662.

79. "Praise for Paris Holocaust Art Exhibition," *The Jewish Chronicle*, February 11, 2010.

80. Christian Boltanski's *Prende la Parole* was a multisensory exhibition shown at Marian Goodman Gallery from September 3 to October 15, 2005.

81. Christian Boltanski and Ralf Beil, *Boltanski: Time* (Darmstadt: Hatje Cantz, 2006), 62.

82. Sarah Rosenbaum-Kranson, "Christian Boltanski," *Museo Magazine*, 2010, http://museomagazine.com/CHRISTIAN-BOLTANSKI.

83. Garb, "In Conversation," 19.

84. Maerkle, "Memory, Too, Fails at Such Excess."

85. Daniel Demetriou, "Boltanski's Heart Don't Skip A Beat," *The Japan Times*, August 6, 2010, https://japantimes.co.jp/culture/2010/08/06/arts/boltanskis-hearts-dont-skip-a-beat/#.XrizPxPok6U.

86. Stephanie Bui, "Our interview with Christian Boltanski," *Crash*, 2015, https://crash.fr/our-interview-with-christian-boltanski.

87. Christian Boltanski, *Christian Boltanski* (London: Phaidon, 1997), 126.

88. Sabine Mirlesse, "On Life, Death, God and Working Very Little: ASX in Conversation with Christian Boltanski," *ASX*, 2015, https://americansuburbx.com/2015/12/on-life-death-god-and-working-very-little-asx-in-conversation-with-christian-boltanski-2015.html.

89. Mendelsohn, *Christian Boltanski*, 158.

90. Philippe Ariès, *Western Attitudes Toward Death: From the Middle Ages to the Present* (Baltimore, MD: Johns Hopkins University Press, 1974), 106.

91. Jean-Luc Nancy, "L'Intrus," *CR: The New Centennial Review* 2, no. 3 (2002): 6.

92. Mendelsohn, *Christian Boltanski*, 159.

93. Dorothy Spears, "Exploring Mortality with Clothes and a Claw," *New York Times*, May 9, 2010, https://nytimes.com/2010/05/10/arts/design/10boltanski.html.

94. Catherine Grenier, *Christian Boltanski: There's a Story* (Paris: Flammarion, 2010), 96.

95. Walter Benjamin, "The Work of Art in the Mechanical Reproduction," in *Illuminations, ed. Hannah Arendt, trans. Harry Zohn* (New York: Schocken Books, 1969), 7.

96. Boltanski and Beil, *Boltanski: Time*, 57.

97. Boltanski and Beil, 65.

INDEX

Pages in italics refer to illustrations.

Abramović, Marina, 232
Absence, 168, 180, 215, 239, 242, 247
Afrofuturism, 81, 84
And She Had a Heart! (Anatomy of the Heart) (Simonet), 23, *25*
Anémic Cinéma (Duchamp), 170
Antiochus and Stratonice (Ingres), 22
Anthropometry, 53–55, 185, 190–191. *See also* Biometrics
Aranda, Julieta, 142–150
Aranda, Luis Jiménez, 26, *27*
Archives, 241, 243–247
Aristotle, 2, 23
Armstrong, David, 200, 201
Armstrong, Neil, 74, 92
Arrythmia, 144–146
Artaud, Antonin, 122–123
Artificial heart, 49–52, 163, 171–178
Artificial intelligence, 3, 182, 201, 213
Auenbrugger, Leopold, 26–28
Auroras, 70–71, 79, 93, 171
Auscultation, 26–31, 33–38, 219, 222
  mediate auscultation, 31
Avicenna, 19–20, 32
Ayurvedic medicine, 1, 14–15
Ayurvedic Practitioner Taking the Pulse, An (unknown), 14, *15*

"Baby's Heartbeat" (Ono and Lennon), 226
Bachelard, Gaston, 107
Balla, Giacomo, 53, *57*
Barnard, Christiaan, 159–160, 162, 163
Beat Piece (Ono), 218, 221
Bell, Jocelyn, 79
Bertillon, Alphonse, 53–55, 181, 190
Biofeedback, 134–136
Biogenesis (Radigue), 70, 93–98
Biomedicine, 3, 8, 73, 119, 131–132, 136
Biometrics, 53–54, 181–182, 185, 191–192, 202, 206–215
Biopolitics, 23, 148, 162, 178–181, 201
Birke, Lynda, 8, 197
Blair, Kirstie, 7–8, 37
Blood
  circulation, 2–3, 20, 41–42, 62, 236, 269n22
  and identity, 163, 192, 211
  as material, 117–120, 132–133, 219
  pressure, 41–42, 45, 52
  visualization, 151, 153, 156–158
Blood Piece (Ono), 219
Body Is a Big Place, The (Pynor), 176–177
Boethius, 3, 32, 222
Boltanski, Christian, 38, 239–247
Border Tuner (Lozano-Hemmer), 209–211
Bowie, David, 83
Boyle, Mark, 117–126, *124*, 193
Boyle Family, 119. *See also* Boyle, Mark; Hills, Joan
Brain, Robert, 7, 53
Broken-heart syndrome, 235
Bruscky, Paulo, 129
Buddhism, 97, 220. *See also* Zen Buddhism
Bulgakov, Mikhail, 65
Burga, Teresa, 183–193

Cage, John, 162, 220, 232, 265n19
Canon of Medicine (Avicenna), 19–20, *21*
Cappa, Benedetta, 65–67
Cardiogram for Africa (Mack), 104, 107–108

Cardiogram of My Heart (Mack), 100, *101*, 103
Cardiogram of the Cyclopes (Mack), 100, 103
Cardiograph, 45, 166
Chakra, 236–237
Charged Hearts (Richards), 171–173, *172*, 177
Chi. *See* Qi
Chi kung, 87, 236
Chinese medicine, 1–2, 7, 14–17, 81, 221–222
Chorus for Six Hearts (Dupuy), 115–117, *116*, 266n34
Chronophotography, 4, 47–49, 53, 57, 169
Circadian rhythm, 107, 137, 148, 224, 269n22
Clar, Richard, 92
Clepsydra, 18, 25
Clock, 67, 142–144, 147, 148, 224–225
  diagnosis, 18, 23
  measurement, 25–26, 137
Closeness
  emotional, 24, 78, 140–141, 223
  physical, 30–31, 199, 222–223
Co-mingling (Richards), 178–180, *179*
Cone Pyramid (Dupuy). *See Heart Beats Dust: Cone Pyramid* (Dupuy)
Connect—Heartbeat Tokyo (Sasaki), *153*, 156
Corps étranger (Hatoum), 193–194, *196*, 197, *198*, 199–200
Corvisant, Jean Nicolas, 28–30
Crawford, Kate, 201, 213
Cuevas, José Luis, 189
Cubism, 53, 56–57
"Cyborg Manifesto, A" (Haraway), 194, 212
Cyborgs, 173, 212

Dalí, Salvador, 67
Dance, 5, 32, 81, 115, 134
  description of heartbeat, 33
  sculpture, 109–113
Davies, Char, 134
Death, 159–160, 176–180, 204, 232, 237, 241–247
Deep Listening (Oliveros), 87, 236
Derrida, Jacques, 90, 164, 223
Diagnosis, 18, 28–31, 35–36, 44
Doctor Taking a Woman's Pulse (unknown), 20, *21*
Domingues, Diana, 126–136, *128*, *131*, *135*
Downey, Juan, 188
Drawing, 104, 107–108, 151–157, 183, 185
Drumbeats, 26–28, 32, 81–84, 133, 136, 228–229
Druyan, Ann, 76, 78
Duchamp, Marcel, 53, *56*, 113, 160–165, 167–171
Duchamp Boxed (O'Doherty), 168, *169*
Dupuy, Jean, 38, 108–111, *112*, 117

E.A.T. *See* Experiments in Art and Technology
Echoes from the Moon (Oliveros), 89–91
Egyptian medicine, 1, 17–18, 28
Einthoven, Willem, 4, 58, 62, 64
EKG. *See* Electrocardiography
Electrical action of the heart, 62
Electrocardiogram, 13, 60, 63–64, 104, 164, 186, 192
Electrocardiography, 4, 58–64, 129, 142–145, 160–171, 189
Electromagnetism, 10, 58–64, 132, 171–177
  bioelectromagnetics, 47
Electropop, 84
"Elektro Kardiogramm" (Kraftwerk), 84, *85*
Eletrourbs, Eletrogravuras (Domingues), 127–129, *128*
Emotions, 2–3, 8, 66–67, 123–126, 172–174, 226–238
  hot hearts, 23
Endurance art, 232–233, 277n46
Energy. *See also* Electromagnetism
  bodily, 119, 123, 127, 134, 149, 159
  electrical, 1, 58–62, 65, 99–100, 127, 132–136, 177
  libidinal, 120, 123

movement, 47–52, 65, 103
optical, 103–104
psychic, 23
spiritual, 132 (*see also* Prana; Qi)
trace, 1, 41, 62, 104–106
vibrational, 102–108, 151
visual, 7, 115–116
Energy Changes (Oliveros), 87–88
Ephémère (Davies), 134
Erasistratus, 18–19, 23
Ernst, Max, 57–58
Eternal Time (Ono), 224, *225*
Examination, The (Faivre), 28–30, *29*
Experimental music, 81, 85–98, 122, 219–220, 228
Experiments in Art and Technology, 109–111

Faivre, Jules-Abel, 28–30, *29*
Feminism, 8, 85, 186, 189, 238
15 Days, Heartbeat Drawing (Sasaki), 156
Floyer, John, 25, 137
Fludd, Robert, 33, 44
Fluxus, 129, 220, 221, 224
Foucault, Michel, 23, 191, 200
Frankenstein (Shelley), 62, 163, 176
Freud, Sigmund, 23
Futurism, 65–67, 149

Gagarin, Yuri, 74
Galen, 8, 32, 41, 61, 235
Galvani, Aloysius Luigi, 61–62
Gaze, 197–200
Gender
construction, 8, 30, 119, 162, 200
division, 23, 30–31, 119, 126, 178, 218
Giant Step (Clar), 92
Golden Record, 76–80, *77*, 93
Goldstein, Jack, 232
Graves, Milford, 81, 262n31
Greek medicine, 18–19, 235. *See also* Galen

Haptic method, 17–18
Haraway, Donna, 8, 173, 192, 194, 200, 212, 233
Harvey, William, 20, 41, *42*, 45, 64
Hatoum, Mona, 193–201
Hayles, N. Katherine, 130, 213
Healing, 27–28, 32, 81, 123, 132–134, 234–239
Heart Beating Machine (Burga), 185–186
Heartbeats (Fujimura), 206
Heart Beats Dust: Cone Pyramid (Dupuy), 108–113, *110*, *112*
Heartbeats/Syria (Ono), 229, 241
Heartbeat stars, 79, *80*
Heartbeat Venice (Sasaki), *154*, *155*, 156–157
Heartbreak, 5, 174, 233–235
Heartlight (Borofsky), 206
Heart murmur, 3, 230, 233–235
Heart Murmur (Montano), 230, *231*, 233–235
Heart physiology, 1–3, 19–20
Heart Rhythm (Mack), 104, *105*
Heartscapes (Domingues), 134–136, *135*
Henry, Pierre, 95–96
Herophilus of Alexandria, 18–19, 25, 32, 137
Hills, Joan, 117–126, *124*, 193
Hospital Ward during the Chief Physician's Round, A (Aranda), 26, *27*
Humors, 8, 14

ibn Sina, Abu Ali. *See* Avicenna
Identification systems, 3, 181–182, 185, 191–192, 212–214. *See also* Biometrics
Indefinite Integral of Psi Star Psi D Tau Equals One (Oliveros), 88–89, 91
Index, 140, 166, 189, 213–214
Indigenous peoples. *See also* Medicine, traditional; Shamanism
Aboriginal Australians, 32
First Nations (Canada), 32
Kuikuro, 132–133
Nahuas, 2, 28
Maya, 2,
Māori, 32
Information, 182, 189–193, 201, 211–214
Interaction, 96, 109–111, 134, 151–152, 177, 209

Irigaray, Luce, 200, 238
It will, it will. I've guaranteed it (Aranda), 144–146, *146*

"John & Yoko" (Ono and Lennon), 228

Kahlo, Frida, 236
Kahn, Douglas, 8, 89
Kaprow, Allan, 138–142
Khlebnikov, Velimir, 149
Kinetic art, 53, 68, 100–103, 123, 164
Klüver, Billy, 111
Kraftwerk, 84, *85*
Kuriyama, Shigehisa, 7, 221
Kymograph, 42

Laennec, René, 4, 30–31, 33–37, *34*, 114, 222
Laennec Listening to the Chest of a Patient (Board), 34
Laika, 71–73, 84
La Jetée (Marker), 150
Lang, Fritz, 66–67
Le Coeur (Boltanski), 242–243
Lefebvre, Henri, 107
Lennon, John, 218, 226–229
Leonardo da Vinci, 20, 33, *40*, 41
Les Archives du Coeur (Boltanski), 241, 243–246, *244*, *245*
Letters from Linda M. Montano (Montano), 237
Lighthouse (Fishkin), 206
"L'intrus" (Nancy), 174–176
L'intrus (after Charged Hearts) (Richards), 174–177, *175*
Little Diagram of My Heart (Mack), 103
Llano-Mejía, Sandra, 189
Look No Body! (Hatoum), 194–195, 199
Lovesickness, 22–23, 65, 255n17
Lozano-Hemmer Rafael, 202–215, 241
Lucier, Alvin, 91–92, 266n34

Macario (Gavaldón), 204, *205*, 211
Mack, Heinz, 100–109, 151, 189
Mandarin Doctor Consulting a Patient, A (Zhou), 14, *16*
Marey, Étienne-Jules, 4, *39*, 39, 41, 43–48
Marinetti, Filippo Tommaso, 65–67
Marker, Chris, 150
Mazrui, Ali, 162–163
Meat Joy (Schneemann), 123
Mediate auscultation. *See under* Auscultation
Medicine, 7–8, 20–26, 33–39, 160–164
  instruments, 26–32, 129–132, 200–201, 212
  listening (*See* Auscultation)
  traditional, 1–2, 7, 14–20, 32, 81
Memory, 150, 226, 240–241, 243–244, 247
Metropolis (Lang), *66*, 67
Mikami, Seiko, 151
Mobile Feelings (Sommerer and Mignonneau), 174
Moeller, Christian, 206
Montano, Linda, 38, 229–239
Moon Bounce (Lucier), 91–92
Morris, Robert, 167
Movement (Marey), 48
Musique concrète, 95

Nancy, Jean-Luc, 168, 174–176, 177, 185, 223, 242, 248
Nollet, Jean-Antoine, 60, 206
Nude Descending a Staircase No. 2 (Duchamp), 53, *56*, 169

O'Doherty, Brian, 160–171, 189
Oliveros, Pauline, 85–93, *86*, 230, 232–234, 236
100,000-beats / 24 hours (Sasaki), 157
Ono, Yoko, 7, 152, 218–229, 241
On Pulse (Herophilus), 18

Palpation, 14–22, 45, 138
Papyrus Ebers, 18, 28
Personnes (Boltanski), 239, *240*, 242
Phonocardiography, 38, 183, 226
Physician's Visit, The (Steen), 23, *24*
Pink Floyd, 83, 122
Piscator, Erwin, 67
Pneuma, 18–19

Poe, Edgar Allan, 37, 167
"Poetics of a Transplanted Heart, The" (Mazrui), 162–163, 269n5
"Politics and Esthetics of Heart Transplants, The" (O'Doherty), 163
Pollock, Anne, 8, 226, 276n30
Polygraph, 39, 45, 47, 207
Portrait of Marcel Duchamp series (O'Doherty), 160–164, *161*, *165*, 168, 170–171
Prana, 1
Presence, 168, 180, 215, 218, 239, 242, 244–247
Psychedelia, 83, 120–122
Pulsação Ou Onda Ou Vermelho, Não Verde (Domingues), 129
Pulsars, 76, 78, 79
Pulse
  detection, 14–22 (*see also* Palpation)
  diagnosis, 14–19
  as language, 23, 137, 221
  musical interpretations of, 32–35, 52
  as sign, 2–6, 11, 68, 137, 140, 207
  types, 16–17
Pulse (Dupuy), 117
Pulse Corniche (Lozano-Hemmer), 207–209, *208*
Pulse Front: Relational Architecture 12 (Lozano-Hemmer), 207–209
Pulse Index (Lozano-Hemmer), 213–214, *214*
Pulse Phone (Lozano-Hemmer), 211–212
Pulse Piece (Ono), 218, 221
Pulse Room (Lozano-Hemmer), 202–204, *203*, 209
Pulse Topology (Lozano-Hemmer), 202, 204
Pulse-Watch, The (Floyer), 25
Pulsus (Fludd), 33, 44
Pynor, Helen, 176–177

Qi, 14, 87

Radigue, Éliane, 38, 93–98, 209, 250
Remote Pulse (Lozano-Hemmer), 209–211
Rest Energy (Abramović and Ulay), 232
Rhythm, 162–163, 166–168, 176–178. *See also* Pulse
Rhythmanalysis (Lefebvre), 107
Rhythm of Seeing (Mack), 104–107, *106*
Richards, Catherine, 171–180
Ritual, 127–136, 230, 236
Ritual Meal (Smith), 236
Rosler, Martha, 190
Rotating Brains/Beating Heart (Oliveros, Stelarc, Pearson), 90–91
Royal Heart, The (Dalí), 67

Sacred Heart of Jesus, 5, 102–103
Sagan, Carl, 76, 78
Sasaki, 151–157
Schaeffer, Pierre, 95–96
Schneemann, Carolee, 123
Science of the Heart (Viola), 197
Scores, 35, 88, 218–223
Self-Portrait. Structure. Report. 9.6.1972 (Burga), 182–183, 185–188, *186*, *188*, 191, 193
Sensual Laboratory, 119, 120. *See also* Boyle, Mark; Hills, Joan
Sentimentos: um poema feito com o coração (Bruscky), 129
7 Days, August–October 2000 (Sasaki), 156
7 Years of Living Art (Montano), 236–237
Shamanism, 28, 132–136
Shiota, Chiharu, 152
Sin título (Heartbeat Recording and Light) (Burga), 183–185, *184*
Slowly/mulled over/matter (Aranda), 144, *145*
Smith, Barbara T., 232, 236
Sommerer, Christa, 174
Son et Lumière: Bodily Fluids and Functions (Boyle and Hills), 117–126, *121*, *124*
Sonic Meditations (Oliveros), 87, 236
Sonosphere, 89, 92–93
Space flight, 71–75, 89, 92, 150
Sphygmograph, 4, 13, 42–47, 58, 64
Stealing One's Own Corpse (Aranda), 150

Steen, Jan, 23, *24*
Stelarc, 90–91, 206
Sterne, Jonathan, 7, 31, 33, 256n37
Stethoscope, 4, 13, 26–31, 33–38, 45, 114, 229–234
Struthius, Josephus, 33, 41
Study of a Girl Running on a Balcony (Balla), 53, *57*
Summers, Elaine, 87
Sun Ra, 69, 81, *82*, 89
Surrealism, 57–58, 67, 187

Techno, 84
Technoscience, 4–9, 131–132, 181, 251
"Tell-Tale Heart, The" (Poe), 37, 167
Time/Bank (Aranda and Vidokle), 149
Time Divisa (Macotela), 149
Time Pieces (Kaprow), 138–141
Tosa, Naoko, 134
Touch Poem for Group of People (Ono), 222
Trance, 127, 132–134
Trans-E: My Body, My Blood (Domingues), 126–127, 132–133
Trans-E: The Body and Technologies (Domingues), 130–131, *131*
Two shakes, a tick, and a jiffy (Aranda), 142–144, *143*, 147–148

Ulay, 232, 277n48
Ultrasound, 130, 132, 204
Unfinished Music No. 2: Life with the Lions (Ono and Lennon), 226, *227*
Untitled (Hook, Track, Memory Dents) (Morris), 167

van der Ploeg, Irma, 201, 213
Vibration. *See* Energy
Victorian Poetry and the Culture of the Heart (Blair), 7–8, 37
Vierordt, Karl von, 42–43
Viola, Bill, 197
Virtual reality, 90–91, 130, 134–136, 173
Visual Energy of Sound, The (Dupuy), 115
Volta, Alessandro, 61–62
Voyager Golden Record. *See* Golden Record

Watches, 25–26, 137–138, 212
Waves
  electromagnetic, 13, 79–80, 99, 132–134, 171–173
  pulse as, 20, 41–43, 213
  waveforms, 13, 41, 63, 123, 129, 164–166, 177
World, Membrane and the Dismembered Body (Mikami), 151

Zen Buddhism, 81, 87–88, 138, 140, 220–223, 230–232
ZERO, 100–102
Zhou Pei Qun, 14, *16*